Ethical Challenges in the Behavioral Brain Sciences

In recent years, a growing number of scientific careers have been brought down by scientists' failure to satisfactorily confront ethical challenges. Scientists need to learn early on what constitutes acceptable ethical behavior in their professions. *Ethical Challenges in the Behavioral and Brain Sciences* encourages readers to engage in discussions of the diverse ethical challenges encountered by behavioral and brain scientists. The goal is to allow scientists to reflect on ethical issues before potentially confronting them. Each chapter is authored by a prominent scientist in the field, who describes a dilemma, how it was resolved, and what the scientist would do differently if confronted with the situation again.

Featuring commentary throughout and a culmination of opinions and experiences shared by leaders in the field, this book has as its goal not to provide "correct" answers to real-world ethical challenges; instead, authors pose the challenges, discuss their experiences and viewpoints on them, and speculate on alternative reactions to the issues. The firsthand insights shared throughout the book will provide an important basis for reflection among students and professionals on how to resolve the kinds of ethical challenges they may face in their own careers.

Robert J. Sternberg is past president of the Federation of Associations in Behavioral and Brain Sciences. He has been a professor at Yale University, Tufts University, Oklahoma State University, and the University of Wyoming and is currently a professor at Cornell University. He is a member of the American Academy of Arts and Sciences and the National Academy of Education and a former president of the American Psychological Association.

Susan T. Fiske is president of the Federation of Associations in Behavioral and Brain Sciences. She has been a professor at Carnegie Mellon University and the University of Massachusetts Amherst and is currently a professor at Princeton University. She is a member of the American Academy of Arts and Sciences and the National Academy of Sciences and a former president of the Association for Psychological Science.

Ethical Challenges in the Behavioral and Brain Sciences

Case Studies and Commentaries

Edited by

Robert J. Sternberg
Cornell University

Susan T. Fiske
Princeton University

CAMBRIDGE
UNIVERSITY PRESS

CAMBRIDGE
UNIVERSITY PRESS

32 Avenue of the Americas, New York, NY 10013-2473, USA

Cambridge University Press is part of the University of Cambridge.

It furthers the University's mission by disseminating knowledge in the pursuit of education, learning, and research at the highest international levels of excellence.

www.cambridge.org
Information on this title: www.cambridge.org/9781107671706

© Cambridge University Press 2015

This publication is in copyright. Subject to statutory exception and to the provisions of relevant collective licensing agreements, no reproduction of any part may take place without the written permission of Cambridge University Press.

First published 2015

Printed in Great Britain by Clays Ltd, St Ives plc

A catalog record for this publication is available from the British Library.

Library of Congress Cataloging in Publication data
Ethical challenges in the behavioral and brain sciences : case studies and commentaries / editors, Robert J. Sternberg, Cornell University, Susan T. Fiske, Princeton University.
    pages cm
Includes bibliographical references and index.
ISBN 978-1-107-03973-5 (hardback) – ISBN 978-1-107-67170-6 (paperback)
1. Psychology – Moral and ethical aspects.   I. Sternberg, Robert J.
II. Fiske, Susan T.
BF76.4.E8184   2014
174'.915–dc23   2014020945

ISBN 978-1-107-03973-5 Hardback
ISBN 978-1-107-67170-6 Paperback

Cambridge University Press has no responsibility for the persistence or accuracy of URLs for external or third-party Internet Web sites referred to in this publication and does not guarantee that any content on such Web sites is, or will remain, accurate or appropriate.

Contents

List of Contributors	*page* xi
Preface	xv

Part I Academic Cheating

1 Beyond the Immediate: Academic Dishonesty   3
RICHARD ABRAMS

2 Collaboration, Cheating, or Both?   5
JANETTE B. BENSON

3 Grappling with Student Plagiarism   8
SCOTT PLOUS

4 Commentary to Part I   11
SUSAN T. FISKE

Part II Academic Excuses and Fairness

5 The Compassionate Instructor Doesn't Always Award Extra Credit   15
WILLIAM BUSKIST

6 An Ethical Dilemma in Teaching   18
EVA DREIKURS FERGUSON

7 Attempted Retribution by a Disgruntled Individual   20
JOHN HAGEN

8 Grading and the "Fairness Doctrine"   22
JAMES S. NAIRNE

9 Managing and Responding to Requests by Students Seeking to Improve Their Achievement-Related Outcomes   25
SHARON NELSON-LE GALL AND ELAINE F. JONES

10 Are There Times When Something Is of Greater Importance Than the Truth? 28
BERNARD WEINER

11 Commentary to Part II 31
SUSAN T. FISKE

Part III Authorship and Credit

12 An Ethical Dilemma in Publishing 35
LARRY E. BEUTLER

13 What Does Authorship Mean? 38
DALE C. FARRAN

14 The Ethical Use of Published Scales 41
DIANE F. HALPERN

15 Idea Poaching Behind the Veil of Blind Peer Review 44
RICK H. HOYLE

16 An Ethical Challenge 48
SUSAN KEMPER

17 Authorship: Credit Where Credit Is Due 50
STEPHEN M. KOSSLYN

18 Publication of Student Data When the Student Cannot Be Contacted 53
PETER F. LOVIBOND

19 Ethics in Research: Interactions between Junior and Senior Scientists 55
GRETA B. RAGLAN, JAY SCHULKIN, AND ANONYMOUS

20 Resolving Ethical Lapses in the Non-Publication of Dissertations 59
MICHAEL C. ROBERTS, SARAH E. BEALS-ERICKSON, SPENCER C. EVANS, CATHLEEN ODAR, AND KIMBERLY S. CANTER

21 Theft 63
NAOMI WEISSTEIN

22 Claiming the Ownership of Someone Else's Idea 66
DAN ZAKAY

23 Commentary to Part III 68
SUSAN T. FISKE

Part IV Confidentiality's Limits

24 Ethics in Service   73
ROBERT PRENTKY

25 Protecting Confidentiality in a Study of Adolescents' Digital Communication   76
MARION K. UNDERWOOD

26 Commentary to Part IV   80
SUSAN T. FISKE

Part V Data Analysis, Reporting, and Sharing

27 Clawing Back a Promising Paper   83
TERESA M. AMABILE, REGINA CONTI, AND HEATHER COON

28 When the Data and Theory Don't Match   85
BERTRAM GAWRONSKI

29 Desperate Data Analysis by a Desperate Job Candidate   87
JONATHAN HAIDT

30 Own Your Errors   89
DAVID HAMBRICK

31 Caution in Data Sharing   91
RICHARD L. MORELAND

32 The Conflict Entailed in Using a Post Hoc Theory to Organize a Research Report   94
THOMAS S. WALLSTEN

33 Commentary to Part V   98
SUSAN T. FISKE

Part VI Designing Research

34 Complete or Incomplete, That Is the Question: An Ethics Adventure in Experimental Design   101
NANCY K. DESS

35 "Getting It Right" Can Also Be Wrong   105
RONNIE JANOFF-BULMAN

36 Commentary to Part VI   108
SUSAN T. FISKE

Part VII Fabricating Data

37 Beware the Serial Collaborator 111
DAVID C. GEARY

38 My Ethical Dilemma 114
SCOTT O. LILIENFELD

39 Data Not to Trust 119
DANIELLE S. MCNAMARA

40 When a Research Assistant (Maybe) Fabricates Data 121
STEVEN L. NEUBERG

41 The Pattern in the Data 124
TODD K. SHACKELFORD

42 It Is Never as Simple as It Seems: The Wide-Ranging Impacts of Ethics Violations 126
MICHAEL STRUBE

43 Commentary to Part VII 128
SUSAN T. FISKE

Part VIII Human Subjects

44 Ethical Considerations When Conducting Research on Children's Eyewitness Abilities 131
KYNDRA C. CLEVELAND AND JODI A. QUAS

45 Studying Harm-Doing without Doing Harm: The Case of the BBC Prison Study, the Stanford Prison Experiment, and the Role-Conformity Model of Tyranny 134
S. ALEXANDER HASLAM, STEPHEN D. REICHER, AND MARK R. MCDERMOTT

46 Observational Research, Prediction, and Ethics: An Early-Career Dilemma 140
STEPHEN P. HINSHAW

47 Should We Tell the Parents? Balancing Science and Children's Needs in a Longitudinal Study 145
KATHY HIRSH-PASEK AND MARSHA WEINRAUB

48 Ethics in Human Subjects Research in Brazil: Working with Victims of Sexual Violence 149
SILVIA H. KOLLER AND LUISA F. HABIGZANG

49	Honesty in Scientific Study WILLIAM B. SWANN	153
50	Ethically Questionable Research WILLIAM VON HIPPEL	155
51	Commentary to Part VIII SUSAN T. FISKE	157

Part IX Personnel Decisions

52	Culture, Fellowship Opportunities, and Ethical Issues for Decision Makers RICHARD W. BRISLIN AND VALERIE ROSENBLATT	161
53	Balancing Profession with Ego: The Frailty of Tenure Decisions P. CHRISTOPHER EARLEY	165
54	Fidelity and Responsibility in Leadership: What Should We Expect (of Ourselves)? DONALD J. FOSS	167
55	To Thine Own Self Be True DAVID TRAFIMOW	171
56	When Things Go Bad ROBERT J. VALLERAND	174
57	Commentary to Part IX SUSAN T. FISKE	177

Part X Reviewing and Editing

58	The Ethics of Repeat Reviewing of Journal Manuscripts SUSAN T. FISKE	181
59	Bias in the Review Process JOAN G. MILLER	183
60	The Rind et al. Affair: Later Reflections KENNETH J. SHER	186
61	Me, Myself, and a Third Party STEVEN K. SHEVELL	191
62	Commentary to Part X SUSAN T. FISKE	194

Part XI Science for Hire and Conflict of Interest

63 The Power of Industry (Money) in Influencing Science 197
K. D. BROWNELL

64 The Impact of Personal Expectations and Biases in
Preparing Expert Testimony 200
RAY BULL

65 The Fragility of Truth in Expert Testimony 202
PHOEBE C. ELLSWORTH

66 A Surprising Request from a Grant Monitor 205
ROBERT J. STERNBERG

67 Whoever Pays the Piper Calls the Tune: A Case of
Documenting Funding Sources 208
HOWARD TENNEN

68 How to Protect Scientific Integrity under Social and
Political Pressure: Applied Day-Care Research between
Science and Policy 212
MARINUS H. VAN IJZENDOORN AND HARRIET VERMEER

69 Commentary to Part XI 217
SUSAN T. FISKE

*Epilogue: Why Is Ethical Behavior Challenging? A Model of
Ethical Reasoning* 219
ROBERT J. STERNBERG

Index 227

Contributors

RICHARD ABRAMS, Washington University in St. Louis
TERESA M. AMABILE, Harvard Business School
SARAH E. BEALS-ERICKSON, University of Kansas
JANETTE B. BENSON, University of Denver
LARRY E. BEUTLER, Palo Alto University
RICHARD W. BRISLIN, University of Hawai'i at Mānoa
K. D. BROWNELL, Duke University
RAY BULL, University of Leicester
WILLIAM BUSKIST, Auburn University
KIMBERLY S. CANTER, University of Kansas
KYNDRA C. CLEVELAND, University of California, Irvine
REGINA CONTI, Colgate University
HEATHER COON, North Central College
NANCY K. DESS, Occidental College
EVA DREIKURS FERGUSON, Southern Illinois University
P. CHRISTOPHER EARLEY, Purdue University
PHOEBE C. ELLSWORTH, The University of Michigan Law School
SPENCER C. EVANS, University of Kansas
DALE C. FARRAN, Vanderbilt University
SUSAN T. FISKE, Princeton University
DONALD J. FOSS, University of Houston
BERTRAM GAWRONSKI, University of Texas at Austin
DAVID C. GEARY, University of Missouri
LUISA F. HABIGZANG, Universidade Federal do Rio Grande do Sul, Brazil
JOHN HAGEN, University of Michigan
JONATHAN HAIDT, NYU Stern School of Business
DIANE F. HALPERN, Claremont McKenna College
DAVID HAMBRICK, Michigan State University
S. ALEXANDER HASLAM, University of Exeter
STEPHEN P. HINSHAW, University of California, Berkeley

KATHY HIRSH-PASEK, Temple University
RICK H. HOYLE, Duke University
RONNIE JANOFF-BULMAN, University of Massachusetts, Amherst
ELAINE F. JONES, University of Pittsburgh
SUSAN KEMPER, University of Kansas
SILVIA H. KOLLER, Universidade Federal do Rio Grande do Sul, Brazil
STEPHEN M. KOSSLYN, Minerva Project
SCOTT O. LILIENFELD, Emory University
PETER F. LOVIBOND, University of New South Wales, Sydney
MARK R. MCDERMOTT, University of East London
DANIELLE S. MCNAMARA, Arizona State University
JOAN G. MILLER, New School for Social Research
RICHARD L. MORELAND, University of Pittsburgh
JAMES S. NAIRNE, Purdue University
SHARON NELSON-LE GALL, University of Pittsburgh
STEVEN L. NEUBERG, Arizona State University
CATHLEEN ODAR, University of Kansas
SCOTT PLOUS, Wesleyan University
ROBERT PRENTKY, Fairleigh Dickinson University
JODI A. QUAS, University of California, Irvine
GRETA B. RAGLAN, American Congress of Obstetricians and Gynecologists
STEPHEN D. REICHER, University of St. Andrews
MICHAEL C. ROBERTS, University of Kansas
VALERIE ROSENBLATT, San Francisco State University
JAY SCHULKIN, Georgetown University
TODD K. SHACKELFORD, Oakland University
KENNETH J. SHER, University of Missouri
STEVEN K. SHEVELL, University of Chicago
ROBERT J. STERNBERG, Cornell University
MICHAEL STRUBE, Washington University in St. Louis
WILLIAM B. SWANN, University of Texas at Austin
HOWARD TENNEN, University of Connecticut Health Center
DAVID TRAFIMOW, New Mexico State University
MARION K. UNDERWOOD, University of Texas at Dallas
ROBERT J. VALLERAND, McGill University
MARINUS H. VAN IJZENDOORN, Leiden University
HARRIET VERMEER, Leiden University
WILLIAM VON HIPPEL, University of Queensland
THOMAS S. WALLSTEN, University of Maryland

BERNARD WEINER, University of California, Los Angeles
MARSHA WEINRAUB, Temple University
NAOMI WEISSTEIN, State University of New York at Buffalo (Emeritus)
DAN ZAKAY, Tel Aviv University

Preface

When we were in graduate school, no one paid a whole lot of attention to ethics – neither in teaching nor in research. There were no courses on ethics in our graduate curricula and no serious informal instruction either. Human-subjects committees were starting to be formed but were viewed as nothing more than unpleasant hurdles to pass through in order to get one's research done. Many behavioral and brain scientists today still view such committees as little more than annoyances. Yet today, ethics looms large for all behavioral and brain scientists. Here's why.

A professor of psychology at one of the top universities in the United States, someone with a previously impeccable reputation, resigned his position following a protracted and painful scandal in which serious questions arose concerning the correct interpretation of his data. Basically, he was accused of reading into the data what he wanted to see in them, regardless of what they said. What is especially puzzling is that almost any psychologist would have been thrilled to have his reputation, or even anything close to it. Why mess with the data?

A professor of social psychology in Europe had become widely famous for his ingenious experiments and his compelling results. He was not only a star in academia but a media darling as well. Today he too is out of a job because it turned out that he not only faked his data but also even faked experiments – claiming to have run experiments that were never executed.

Most ethical lapses are not of the magnitude of these, but less serious ethical lapses are much more common, and start early. Dora Clarke-Pine, an associate professor of psychology at La Sierra University in Riverside, California, conducted a study of psychology PhD dissertations obtained from a national sample of graduate students at both religious and non-religious universities. She found at least one example of plagiarism in four out of five dissertations.

Graduate students are trained in many and diverse aspects of professional conduct. But even today, one area in which training is sparse, and sometimes nonexistent, is professional ethics. This lack of ethical training becomes challenging in an era in which electronic communications make it extremely easy to plagiarize and to cheat in other ways, such as claiming originality for ideas picked up via the Internet from all over the world. Whereas a bad set of results in an empirical paper can spoil the chances of acceptance by a prestigious journal, an ethical lapse in the production of a paper can spoil a career.

The purpose of this book is to educate students and professionals about dealing with ethical challenges in the brain and behavioral sciences through case studies and commentaries. The importance of the case studies cannot be overemphasized. Many individuals receive ethical training of some sort from an early age: from their parents, through religious study, or through specific courses on ethics. The problem, well known in cognitive psychology, is the difficulty of getting transfer. What people learn in an abstract, encapsulated way often is not translated to their everyday behavior. In a world in which even religious leaders make the news for their severe lapses in ethical behavior, it is difficult to find role models and sources of instruction that help guide students and professionals down an ethical path.

Regrettably, there is more pressure than ever before on behavioral and brain scientists to produce exciting papers with compelling results. This pressure comes in several forms. First, the field is greatly expanded in terms of sheer numbers of professionals, meaning greater competition. Second, competition for obtaining grants – for which empirical publications are necessary – is stiffer than ever, with many agencies funding fewer than 10% of submissions. Third, competition for jobs is at record levels. The *Chronicle of Higher Education* (March 23, 2012) reported that aging professors "create a faculty bottleneck" (p. 1). In particular, at some universities, according to the *Chronicle*, 1 in 3 academics are now 60 or older, and the number of professors aged 65 and older has more than doubled between 2000 and 2011 (p. 1). These figures, combined with decreased state support for public institutions, have drastically reduced the number of openings for new faculty members, especially in the junior ranks. Fourth, many of us who teach have found that students just do not have the same ethical standards as they once did. For example, the *Chronicle of Higher Education* (March 12, 2012) has reported that cheating is rampant at British universities, and that more than 45,000 students have been found guilty of academic misconduct at 80 universities over the past 3 years (http://chronicle.com/blogs/global/cheating-is-rife-at-british-universities/32438). Experience in the United States is comparable. Fifth, it is just easier to cheat than ever before. Plagiarism does not even require writing out the text that one is copying without attribution.

One can simply move a block of text from someone else's document to one's own with the click of a mouse. Increasing means of deterrence, such as Turnitin, does not seem to do much to discourage those determined to plagiarize from external sources.

Our sponsoring organization, the Federation of Associations in Behavioral and Brain Sciences, has created through this book a compendium of case studies and commentaries regarding ethical challenges facing scientists in the behavioral and brain sciences.

The editorial board for this book consisted of Max Bazerman, Harvard University; Jenny Crocker, Ohio State University; Susan T. Fiske, Princeton University; Joshua Greene, Harvard University; Todd Heatherton, Dartmouth University; Joseph Simmons, University of Pennsylvania; Uri Simonsohn, University of Pennsylvania; Sam Sommers, Tufts University; and Robert J. Sternberg, Cornell University.

The editors, Sternberg and Fiske, asked behavioral and brain scientists to contribute case studies representing ethical challenges they have faced in their own careers. Contributors were asked to address five issues:

1. A description of the ethical challenge.
2. What, if anything, made solving the ethical challenge difficult.
3. How the scientist resolved the challenge.
4. What the scientist might do differently if he/she were to face the situation today.
5. What general principle, if any, the scientist can infer from the case study.

The case studies they provided are in the pages that follow. We hope you find them useful in your professional work and perhaps even outside it.

Part I

Academic Cheating

1 Beyond the Immediate: Academic Dishonesty

Richard Abrams

It was 1988 and I had been out of graduate school for two years when I encountered my first case of academic dishonesty (at least I had not suspected any dishonesty before that). The course was Experimental Psychology – a laboratory course like those at many universities where the centerpiece of the course is an independent experimental project of the student's own design culminating in the submission of a complete write-up (in APA style, of course) of the experiment. (These days there are PowerPoint presentations in addition to the paper – and a relaxation of the APA style rules.)

A student who had been performing at an average level in the class turned in a report of an experiment on some aspect of memory. (At least I think it was about memory – isn't that what people studied in the 1980s?) The paper was excellent – and that was the problem. How could someone who can write so well, think so clearly, and present results so succinctly receive only a C on my tests, where the biggest challenge is to remember the distinction between a Type I and a Type II error? I knew that something was amiss when one of the dependent variables that he reported revealed a grain of analysis finer than what would be possible with the reported number of participants. He reported the percentage of participants who responded in a particular way, but when converted to a number, the value was not a whole number. In other words, the data had come from a study with a greater number of participants than what he had reported. Eventually I found the article on which his paper was "based."

When I confronted the student and suggested that not only had he not written the paper, but he hadn't even collected the data, he was defensive. And in his defense he provided "proof" that he had written the paper: a printout showing that the file on his computer had been created one month before the end of the semester. Putting aside for the moment the ease with which one might spoof a file creation date (the student argued that since he had an Apple computer that was not possible), I pointed out that even if he had created the file on his computer one month in advance, that did not mean that he hadn't copied the paper. (In retrospect I suppose that the truth is that he had done the whole thing at the last minute and somehow believed that he only needed to convince me that that wasn't the case.)

As a new assistant professor, I consulted with others in my department on the appropriate course of action. The university did have a formal procedure for dealing with cases of academic dishonesty at the time, but it seemed as if the tradition was to handle such cases on a more "personal" individual basis – sending a student off to the judicial board seemed so impersonal, especially at a smallish expensive private university. And so I handled the case myself by assigning a failing grade on the paper (which resulted in his receiving a low but passing grade in the course). A tough break – but perhaps much easier to cope with than a trip to the judicial board. My decision turned out to have been a mistake.

Fast forward to the end of the next semester. Renovation work in the psychology building had just begun, and most of the faculty were avoiding the building when a knock came on my door from an adjunct instructor who was seeking a colleague of mine who was not in the building at the time. The instructor had a primary appointment as an administrator in the dean's office, but had taught a course in the psychology department that semester. Her experience with teaching undergraduates was somewhat limited, so she was seeking out a regular faculty member in the department for some advice: A student had submitted a final paper to her that seemed to be unusually well written for an undergraduate. She was wondering if that was typical of the level of our students. As it turns out, not only was the student the same one that I had dealt with a semester earlier; it was the same paper!

This time formal charges were brought against the student. It was learned that this was not the first time that such charges had been made – there had been a pattern of academic dishonesty that extended back in time (in addition to my experience with the student), a pattern that I had been unaware of because I had handled the situation locally. Eventually it was determined that the student's violations were so egregious that he was expelled from the university. But it was also clear during the proceedings that the same conclusion could have been reached a semester earlier – if I had reported the problem that I had encountered. The lesson: Even in a small university it is not possible for an individual department to see enough of the big picture to appropriately deal with a student's misbehavior. Since that time, when I have had concerns about academic dishonesty (and I should say that the cases are relatively infrequent, and have never been as extreme), I have always brought them to the appropriate college-level committee.

2 Collaboration, Cheating, or Both?

Janette B. Benson

During my first year as an assistant professor, while I was teaching an introductory child development class, two students requested to take a make-up exam. They both presented written documentation for their absence, following the instructor's policies stated in my syllabus. I arranged to have the teaching assistant (TA) proctor their make-up exam. The TA met both students, put them in separate but adjacent rooms, asked that all personal belongings be left outside the testing room, and instructed the students to submit the completed exam to a receptionist when they retrieved their belongings. The TA planned to pick up the completed exams from the receptionist and to score the multiple-choice section before returning both exams to me. The exam consisted of multiple-choice and short-answer questions, plus two essays. Each student wrote a predetermined "codename" on the exam pages so that the exams could be graded blindly, and then exam grades were recorded on a master sheet that linked the codename to the student ID number. After each exam, I would complete an item-analysis of the multiple-choice items to determine which items might be bad (e.g., poorly written, confusing), too easy, or even too difficult (e.g., less than 10% correct response) in order to maintain or discard them from the test bank I used for subsequent exams. The exams were graded by section, not by exam, to ensure consistency in applying the rubric for nonobjective items (e.g., short-answer and essays).

As I was recording the point totals by section for the two make-up exams I noticed that each student received the same total score – 77.5 points out of 100 – although scores for each section varied slightly (e.g., 40 and 41 points out of 49 on the multiple-choice section, respectively, for each student). Then I noticed that of the multiple-choice items that were marked incorrect for each student, 10 items for one student and 11 for the other, they both incorrectly answered the same six items, and each selected the same incorrect response option. I was stunned by this pattern and was suspicious that the two students collaborated, especially since these test scores were between one-half and one full grade higher than their previous exam score, but I wanted to be sure. As a new assistant professor, I had not previously had to deal with student cheating.

I consulted a seasoned departmental colleague who was also a statistician. I explained the situation and my suspicions and asked for his advice. He took one look at the response patterns of the multiple-choice items and said that the probability that each student could have selected the same incorrect response for the same six multiple-choice items without collaboration was extremely small, especially since some of the incorrect response alternatives they selected were not the most frequently incorrect responses selected by the class as a whole, as revealed by the results of the item analysis that I had shared with him. With a very knowing expression on his face he said, "It is very clear they cheated." When I asked him what I should do about it, he said, "Nothing. This probably isn't the first time, it won't be the last time they cheat, and if you pursue this, it will take an inordinate amount of your time. As a young assistant professor, this isn't worth your time. You should be doing your research, publishing, and not dealing with lazy students."

I had very mixed feelings upon receiving this advice. My seasoned colleague was correct about how a young assistant professor should be spending her time, but I also felt strongly that I should not let students get away with cheating as it was unfair to other students in my class, and I took seriously my role as an educator who should also be trying to uphold standards, model ethical behavior, and look for opportunities to make a positive impact.

I invited the students to come together to my office, where I showed them their scored exams and pointed out the remarkable similarities in their response patterns and scores. I then very idealistically asked them to imagine they were me and to tell me what conclusions they might draw from the exams. In my idealism I had hoped that when confronted with the evidence, they would admit their moral downfall, at which point I would meet them halfway to rectify the situation. Their response, almost in unison, was, "It is clear that we studied together, which explains why our responses are so similar." Clearly, this was not how I expected this "teachable moment" to unfold, and I kept hearing the voice of my elder colleague in my ear as my time was being sucked away.

In the meantime, I spoke with the TA and was disappointed to learn that she never returned to check up on the students or to be explicit with the receptionist that the two students should have no contact while taking the exam and when retrieving their personal items when submitting the completed exams. This was an important lesson learned by both the TA and myself to make clear that proctoring an exam also meant observing the test takers. We also noticed these two students stopped sitting together during class, and one would frequently miss class.

In the end, I did not give up on my hopes of a "teachable moment" and spent the time to seek out information about the institution's policies regarding student cheating. I met separately with each student (deploying the "divide and conquer" strategy), and I repeatedly invited each to meet with the department chair if she felt that I was being unfair. One student met with the chair because she was afraid she would be expelled. I finally told each student that she could choose between the following courses of action: (a) I would turn everything over to the University Office of Citizenship and Community Standards and let the issue be resolved through the student judicial process; or (b) the students would admit that the exam results were invalid and that their scores from the first exam, a 67 and 71, would serve as a proxy for the score on the exam in question. In the end, neither student made an outright admission of cheating but chose the second option. The score on the last exam would determine whether the students would earn a passing grade of C– or have to repeat the class, which was required for the major. One student earned enough points on the last exam to pass the course with a C– grade, and the other student earned a solid D–.

At the start of the subsequent academic term, the student who passed the class showed up at my office hours and told me she had thought a lot about what happened, admitted that she had cheated, explained that she was "pressured" by the other student, and felt that she had learned an important lesson, including that she felt she owned me an apology because she wanted to participate actively in her major and did not want to avoid the other classes that I offered. She eventually sought me out for academic advising. The student that did not pass the class did not pass her other classes, was put on academic probation, and subsequently dropped out.

Almost 30 years have passed since this incident, and if faced with the same situation again, I most likely would take the same approach. I never once regretted that I ignored the advice of my elder colleague to save my time and look the other way. I continue to believe that part of my responsibility as a professional is to uphold personal and institutional ethical standards, to model academic ethical values to others, to capitalize on teachable moments, and to provide better advice to younger colleagues than was once offered to me.

3 Grappling with Student Plagiarism

Scott Plous

Some years ago, I was grading final papers for a seminar I teach on the psychology of prejudice, and I noticed an usually eloquent passage from a student who was not an especially strong writer. At first I was impressed with the poetic quality of the passage, but the more I thought about it, the less sure I was that this particular student could have written it, so I searched the web and found that the student had used a professional writer's material without attribution – a clear instance of plagiarism.

In the case study that follows, I'll describe my three-part response and conclude with a few words about why my response fell short of a comprehensive solution.

Part 1: The student. After discovering the act of plagiarism, I promptly emailed the student, pointing out that sections of her paper matched unattributed sources verbatim and constituted plagiarism. I then asked her to email me a list of all passages taken directly from other people's work, along with a citation or web address for each original source. I also wrote that even though I had provisionally given her a score of zero for the paper, I hoped that there was a simple explanation for what I found, and I assured her that I was fully committed to handling the problem as fairly as possible.

In less than an hour, the student replied that she was alarmed by my email message because she believed that she had cited all sources adequately, and because her paper did not, as far as she knew, include any direct quotes from other sources. She then asked me to identify the problematic passages so that we could clear up the matter as soon as possible.

Later that day, I sent her details documenting the plagiarism I had discovered, and I once again asked her to list any other such cases so that I could assess the scope of the problem and understand what went wrong. Then, silence.

After nearly a month had passed, I emailed the student one last request for information, but once again I received no reply, at which point her course grade became permanent. The ultimate result of this episode was that a student whom I had previously admired, and who had seemed destined to earn a grade of A− or B+ in the seminar, instead received a D− and ended the course in disgrace.

Part 2: The instructor. I'll never know whether this case involved an act of deliberate plagiarism, an innocent error, or something in between, but the next time I taught the seminar, I resolved to prevent similar cases from occurring. Perhaps the most important step I took − one that I would encourage all instructors to take − was to address plagiarism explicitly in the seminar syllabus. In my revised syllabus, I included a section referencing the Wesleyan University Honor Code, asked students to abide by the code, and offered these guidelines on writing assignments:

All papers, journal entries, and presentations for this class must be original − not reprinted, excerpted, or adapted from existing work (e.g., papers for other classes, books, articles, web pages). Similarly, any text, tables, figures, or images reproduced from other sources must include clear reference citations, and all quoted passages must use quotation marks to indicate that they are quotations. If you're not sure about how to reference something, please ask me rather than running a risk of violating Wesleyan's Honor Code.

I also discuss these guidelines with students before they write their first paper, and I've posted the revised syllabus online for other instructors to adapt and use as they see fit. Since adding this information to the syllabus six years ago, several students have consulted with me to make sure their citations were appropriate, and to the best of my knowledge, none have committed plagiarism.

Part 3: The institution. A year or two after I changed my seminar syllabus, Wesleyan decided to revise its honor code, and as luck would have it, I was appointed to serve as faculty representative to the task force drafting the revision. At the time, the existing honor code made relatively little distinction between plagiarism and cheating on exams, but after my encounter with plagiarism, I had come to understand that plagiarism and cheating on exams were different enough from each other to warrant separate treatment. Whereas cheating on exams is almost always intentional, plagiarism can be the unwitting result of carelessness or a failure to understand the importance of quotation marks and academic standards of attribution (e.g., that slight paraphrasing does not make borrowed material one's own).

I therefore proposed that the university honor code treat plagiarism and cheating on exams separately, which I'm pleased to say it now does. In addition, Wesleyan now asks faculty advisers to define and discuss plagiarism with incoming first-year students, and the new Honor Code includes a detailed description of plagiarism as:

> the presentation of another person's words, ideas, images, data or research as one's own. Plagiarism is more than lifting a text word-for-word, even from sources in the public domain. Paraphrasing or using any content or terms coined by others without proper acknowledgement also constitutes plagiarism.

Epilogue: Why this solution doesn't entirely solve the problem. My three-part response to plagiarism addressed the problem at the level of the student, instructor, and institution, but it ignored a key ethical issue that takes place at the level of the educational system itself: the greater scrutiny some students face when it comes to the detection of plagiarism. In this particular case, for example, what led me to suspect plagiarism in the first place was that the student in question seemed unlikely to have written such an eloquent passage. If, in contrast, a student with strong writing skills had plagiarized the same passage, I probably would have been impressed and never thought to check whether or not the work was original.

Indeed, the issue isn't simply that students are treated differently, but that students who enter college with weak writing skills and prior educational disadvantages are at greatest risk of being caught plagiarizing. Thus, earlier disadvantages become compounded – disadvantages that may have contributed to my seminar student receiving a course grade of D–, which itself may carry further consequences down the road.

What can be done to treat students more fairly? One answer is to analyze all student work uniformly with plagiarism detection software or services such as Turnitin, but that solution is time consuming, prone to false negatives, and doesn't alter the fact that exceptionally high-quality work will continue to arouse more instructor suspicion when it comes from unexceptional students. A second answer is for instructors to level the playing field by checking exceptional work regardless of who submits it, although that solution is undoubtedly easier said than done. For now, the best I can say is that I continue to grapple with this challenge, and I hope that this brief case study will help other instructors do the same.

4 Commentary to Part I

Susan T. Fiske

We are in a new era of academic cheating, but not necessarily because there is more of it. Rather, cheating 2.0 is all about new technology for Big Data, whether searching databases for papers to plagiarize, checking submitted work for plagiarism, or analyzing tests for patterns of performance.

For faculty, catching cheaters is about system-level monitoring to detect patterns. That is, cheater detection is partly about comparing the individual student's ongoing performance to a sudden, surprising leap in performance, prompting a closer look. Cheater detection on tests in particular is also about comparing two or more students' performance for a surprising similarity in the errors they make (accurate performance being less diagnostic because it is less idiosyncratic). Likewise, while twenty-first-century technology makes plagiarizing easier, mainly thanks to the Internet, it also makes catching students committing plagiarism easier, for the same reason. Recall that just a couple of decades ago, with only physical library resources available to both students and educators, both plagiarising and catching plagiarisers were time- and effort-consuming tasks.

Faculty do engage in system-level monitoring to detect odd patterns, and new technologies help here. But as this part has shown, deciding how to manage cheating is a more human process, creating a teachable moment for the student, the instructor, and the institution.

Part II

Academic Excuses and Fairness

5   The Compassionate Instructor Doesn't Always Award Extra Credit

William Buskist

Without fail, during the final week or so of each semester, I receive a small handful of e-mail from students desperately requesting that I offer them an opportunity to earn more points in the course so that they may improve their final grades. On some occasions, individual students will appear in my doorway during my office hours and make their case personally, often with tears in their eyes. Students generally justify their request by an emotional appeal to my sense of empathy, always implying that I am the only teacher who can save them from impending academic doom of great personal consequence. The most common reasons for their appeals include avoiding (a) being placed on academic probation or suspension, (b) losing an academic scholarship, (c) losing financial support from their parents, or (d) facing the disappointment (and sometimes the wrath) of their parents.

It is no surprise that such strong emotional appeals can be successful in inducing a teacher to help students overcome their dire circumstances. Indeed, it is difficult *not* to feel empathy toward students as they passionately plead their situation, especially in a face-to-face meeting.

When I taught my first college course 35 years ago, several students approached me individually and asked for a chance to earn more points so they could raise their final grade in the course. They each made their personal appeal, and in each case, I asked these students to write a short (4–5-page) essay on a topic of their choice. They all completed the task, and accordingly, I awarded them the points they needed to make the next higher grade in the course.

Then something interesting, in a gut-wrenching way, happened. As I was reviewing my final grade sheet just prior to turning it in to the departmental secretary, I realized that I had treated my class, as a whole, unfairly by not giving all students, regardless of their grade, a chance to earn "extra credit." In truth, there were other students in the course who had borderline grades and would have benefited by having the chance to write the extra-credit essay. My behavior was unethical – perhaps naively so, but unethical nonetheless. I had let myself get caught up in the emotion of the moment, thinking that what I was doing was genuinely helping those students who approached me wanting to do more coursework in exchange for raising their final grade. Rather than helping them, I was actually teaching them that it is perfectly fine not to meet course requirements in a timely manner and that this sort of workarounds was an acceptable means of getting through college.

Too embarrassed about what I had done, I did nothing to rectify the situation. I hoped that nobody would notice, and much to my relief, nobody did. But the shame of my behavior compelled me to consider how to do things differently in the future. I vowed that I would do my best to treat all students fairly and that I would never again let myself be persuaded to change students' grades without a just cause, such as my miscalculation of a grade or a university-approved excuse for missing classwork.

Beginning with the following semester, I devoted a portion of the first day of class to discussing "classroom etiquette," in which I noted that the only ways to earn points in the course were those described in the syllabus and that I would not offer individual students opportunities to raise their grade by completing last-minute papers or other projects. By and large, this tactic has reduced the number of end-of-the-semester student appeals for extra credit beyond those noted in the course syllabus. However, no matter how much I stress the point, there are always a few students each semester who feel they can persuade me to let them write an extra paper for the course.

Over the years, my thinking about how to handle such student pleas for 11th-hour assistance in raising their grade has evolved. Rather than seeing this situation as a problem, I view it as an opportunity to teach students about academic integrity and responsibility. Nowadays, when students send me an e-mail asking that I give them something extra to do so they can accumulate a few more course points toward their final grade, I reply with something like the following, depending on each student's specific circumstances:

Thanks for your note. I very much appreciate your concern about your grade in our course. I am happy to hear that you are willing to do additional work in the course to improve your grade. However, what you are asking me to do goes beyond what I've outlined in the syllabus as a means for earning points in this course. As such, it would be unfair of me to provide you a chance to engage in additional course work without also extending the same opportunity to all of the students in the course. Because it is so late in the semester, offering this opportunity to all students would be a burden to many of your fellow students, especially those who may not have the time to complete another assignment, and to me, who would have to grade all the essays.

I am sorry to learn that your grade in my course is causing such a crisis for you and that I could not offer a viable way of helping you resolve your situation.

Keep in mind, though, that you've had many opportunities to earn points in this course, and for whatever reason, you've not taken advantage of these. Perhaps your situation will give you a good chance to reevaluate your approach to your studies this semester. I will be very happy to meet with you and discuss your study methods, including time management skills. Please free to stop in during my office hours or to e-mail me regarding other possible meeting times should my office hours conflict with your schedule.

The general ethical principle in play here is the fair and equal treatment of all students. Whatever means teachers choose to help students learn the subject matter, they must use methods and activities that are accessible to all students. A secondary but very important principle, also operating in situations like the one I've described, is that teachers have the responsibility to teach their students about fairness, academic integrity, and ethical behavior in addition to course content, regardless of the subject matter. Although teachers often infuse discussions of discipline-specific codes of ethics in their courses, another means of teaching students about ethical behavior is on a student-by-student basis. I have found this approach to be particularly effective because it directly impacts students on a personal and perhaps more poignant level than learning about ethical behavior in the abstract.

6   An Ethical Dilemma in Teaching

Eva Dreikurs Ferguson

Some years ago I had an ethical dilemma regarding classroom teaching. One student was consistently absent from class, and on occasion she missed turning in assignments or taking exams. I discussed this with her, and she explained that she had cancer and was getting treatment at medical facilities and thus had to miss class and assignments. I felt obvious sympathy and asked her to bring medical documents to me as verification so that I could prorate her work. She found reasons on each occasion why she could not bring a document from a medical professional. She brought notes from what were presumed to be family members but not a medical document.

When the end of the semester came she indicated she'd be getting treatment for her cancer during the final exam week and could not take the final when scheduled. She cried and narrated a story that provoked true sympathy. I took her to the department advisor, to discuss the matter with her regarding what possible avenues were available to her so that she'd not flunk the course. Again, she gave a sad story of years of cancer treatment. The advisor and I agreed that if the student brought documentation from the medical facility, the student could get an incomplete and take the exam at a later time. She'd receive a grade for the course after she took the final exam.

I followed her out of the advisor's office, in which she had wept openly about her illness, and on my way to my office I saw her talking with another student in the hall. In that conversation the student was not crying but talking cheerfully and with laughter. That jolted me. The tone of that conversation was a sharp contrast to the tears in the advisor's and in my office just prior.

When the exam week came, the student went to the chair of my department, who was new in the position and not yet a seasoned professor. She presented to him the sad tale of her cancer treatment and showed what appeared to be genuine sadness and distress. She requested that he override my condition that she had to bring documentation from a medical center in order to take the final exam late. He believed her and asked me to justify why I required the documentation. I replied that I was not unsympathetic to the student, but that if she were telling the truth, she'd have no difficulty getting documentation regarding her treatment. I had kept a complete record of the e-mail correspondence between us, so I could verify that I had given her more than enough time to obtain the required documentation. My e-mail records made clear that I had not sprung the requirement of documentation on her, which she claimed in talking with the department chairperson. I faced what appeared to be an ethical dilemma, in requiring documentation for her medical condition when the chairperson was willing to let her take the final exam at "her convenience." I insisted that if she did not bring documentation by the time of the final exam and did not take the final as scheduled, she'd fail the course. She did not bring documentation and did not take the final and did fail the course. Later the chairperson asked me how I knew so certainly that she was likely to be faking her illness, and I told him of the incident of observing her talking with a friend right after her woeful tale with tears to the advisor. Moreover, if she were really getting cancer treatment, she'd have had no problem obtaining documentation to verify the treatment.

If I were in the same situation again, I would have done the same thing.

Possible general principle, especially for young faculty: do not forego evidence in the form of documentation when students want special privileges, and remain true to the need for evidence even when an administrator pressures you to grant the student special privileges.

7 Attempted Retribution by a Disgruntled Individual

John Hagen

Coming from a professor, researcher, and administrator with forty-plus years of experience at a major public university, the case study I provide here is, I believe, unique and raises a number of ethical dilemmas as well as a critique of options for handling such situations. My university is highly competitive, both academically and athletically. As a Division 1 school within the NCAA, it expects approximately 700 student athletes to perform extremely well on the field as well as in the classroom. Part of my research for more than 20 years focused on identifying assets that lead to success in these students. My teaching and administrative duties concerned this and other groups of college students.

A few years ago, I took an administrative position in the department at the invitation of our new chair. In this role I had oversight of the program for more than 1,000 majors in our department. I also continued to teach a course on learning and diversity, which attracted both student athletes and students diverse in terms of race/ethnicity and learning style. Several students each term were certified as having learning differences or disabilities. The course grew in size and popularity and student ratings were always high.

I was not aware at the time that a person unhappy with the administration was not happy with my appointment to the administrative office as well. Our chair heard from the dean's office that someone had launched a complaint about my teaching and my course in particular. The chair did some checking, talked to me, and was satisfied there was no basis for the complaint. Apparently, the disgruntled person then carried the complaints to the provost, who conducted an investigation and concurred with the chair's conclusion. Unbeknownst to us, the individual then contacted a reporter for the local newspaper and made claims that student athletes were being given good grades for minimal work.

The provost called a meeting of relevant parties and presented the findings of the investigation to those present, including two reporters, who up to that time were protecting their source. The provost exposed the source at the meeting and challenged the reporters on their information, which he considered fallacious. However, we then learned that the person had obtained various records of students, especially student athletes, with the help of a secretary, who figured out how to hack information from the records office. One of the reporters then came to my class and tried to obtain statements from students. For the next class, a security guard stood at the door to make sure only students could enter. However, some were still interviewed, after they were told an article was being prepared to pay tribute to 'my "retirement"'.

The university's communications department continued talks with the editor of the newspaper, but we heard nothing else until about three months later when a series of articles appeared. There were interviews as well as transcripts of some students in one article that tried to demonstrate that my course was easy and graded leniently. Other articles in the series contained interviews of other staff and administrators, trying to raise other unrelated criticisms of the athletic department.

There was immediate reaction from the communications office of the university, denying any wrongdoing and providing full support for my course as well as the other persons identified in the articles. I personally received more than 100 statements of support from colleagues, students, and others. The university asked the NCAA to review the claims, and after a brief investigation it issued a statement that there was no verifying documentation for the alleged problems.

The provost asked the university's legal department to intervene, and they employed the assistance of technology experts, who quickly discovered that transcripts had been hacked via the computer in the office of the secretary. This secretary abruptly resigned and moved to another state. She was interviewed, but we do not know what transpired. The provost's office, in consultation with the legal department, concluded that the other person had violated the Family Education Rights and Privacy Act, which protects the privacy of students' educational records. There were meetings this person, his/her legal representation, and the university's legal department. An agreement was reached, and the person resigned.

The case raises many questions, and there was considerable second-guessing by some of my colleagues and administrators concerning the way it was handled. There could have been formal charges, but apparently the university chose not to pursue this course of action, which could have resulted in additional publicity. A final footnote is that the newspaper, which was our city's only paper, suddenly announced it was going out of business. All the staff members were terminated, and it became clear the newspaper had been in deep financial difficulties for the past several years. Attempts at sensational reporting did not stave off the newspaper's ultimate demise.

8 Grading and the "Fairness Doctrine"

James S. Nairne

At most universities instructors are encouraged to state their grading policies in the class syllabus, preferably along with numerical ranges for each grade. Ignoring these guidelines can get you in trouble, especially if a disgruntled student chooses to file a grade appeal. If the student can demonstrate that his or her grade does not match what has been given to other students, or if the grade criteria are not explicitly outlined in the syllabus, then the student has a reasonable chance of winning a grade appeal. Special attention is typically given to a "fairness" criterion – all students in the class should be graded in the same way.

But for those of us who teach large classes, in my case more than 400 students in an introductory psychology course, sticking to the "fairness doctrine" can be a challenge. The final grade is based almost entirely on three or four multiple-choice tests, so poor performance on the first (or any) test can put the student in a hopeless situation. Imagine that I give three 100-point multiple-choice tests and the cutoff for a C is a cumulative 210. Student A flunks the first test, receiving a score of 44, but judiciously comes to office hours and performs well on the second exam (70) and even better on the third (90). Student B never comes to office hours, shows no improvement across tests, but ends up with the same final total (68 + 68 + 68). Do these two students really deserve the same grade?

For Student A, the problem is usually poor metacognitive skills – he or she simply does not understand the nature of testing in large classes, or how to study appropriately. It can be satisfying to see this kind of student develop over the course of the semester; in fact, these are often the students that I get to know best. Here, however, a student who improves to the point of receiving an A− on the final exam still ends up with a D in the course. It is a learning experience for the student, but I am always tempted to jettison the fairness doctrine in such cases and bump the student up the grade distribution. To be honest, this is almost always what I do. Yet my actions are inconsistent with my own syllabus and, consequently, I am at risk of losing an appeal from a vexed Student B.

What can you do? In this particular case you can let students drop the lowest score. But this doesn't solve the problem in principle – you will still end up with students whose scores don't adequately reflect their improvement, or the net knowledge that you think they've gained in the course. You can try to write the syllabus in a way that handles such special cases. Admit that you look for improvement across the course of the semester and bump up those students who are near a grade cutoff. But the concept of "improvement" remains vague, and there is no practical way to quantify personal tutoring received during office hour visits. You could add the following: "Students who do poorly on one of the tests, especially the first one, can raise their grade if they increase their performance on subsequent tests and regularly visit me during office hours for extra help." But with 400-plus students, you are likely to receive dozens of visits from robotic students seeking to satisfy the "office hours" escape clause in the syllabus. What we want to do is reward the self-motivated student, the one who works hard to acquire the tacit knowledge it takes to perform well in large classes.

It's an ethical dilemma without a clear solution – one I face nearly every semester. Should I bend the rules, rewarding those special students whose test scores fail to reflect what they have clearly gained from the course, or do I follow an objective standard and treat everyone the same? To make matters worse, my current practice suffers from a volunteer problem – I am rewarding only those students who personally come to office hours. It is certainly possible that there are other students with a true desire to learn and improve who cannot make the office hours because of work or other commitments. In the end, I simply go with my judgment and adjust the grade when I deem it appropriate. After all, we are being paid for our professional judgment, and not all ingredients that constitute it can be stated clearly in a syllabus. But be ready to accept the consequences – some might accuse you of being unfair, which, in a sense, you are.

Here is another example, one not so easily defended. When I was a young assistant professor, a "Student B" came to see me a few days after the final exam. He had performed consistently poorly throughout the semester, receiving a final grade of D. "I realize I'm getting a D in your class," he said, "and I deserve it. I'm graduating this semester and, well, I figured I could get at least a C without studying. I was wrong. But here's my problem: I've been accepted into medical school in the fall and my acceptance letter stipulates that I cannot get a grade lower than C in my final semester. My entire career hinges on getting a final C in my class. I'm begging you ... please change my grade."

What would you do? I asked him to produce the acceptance letter, which he did, and his characterization of the situation was correct – he could not get a letter grade lower than a C. Obviously he was an excellent student overall (and it was one of the best medical schools in the country), but his one lapse threatened his career. I changed his grade to a C. I cannot justify my action; it certainly violated the fairness doctrine, but I have no regrets. I later discussed my decision with one of the senior professors in my department. This is what he told me: "If you hadn't changed his grade, either the head of the department or the dean would have done it." Once again, my actions were not "fair," in the sense of following the grading contract established by the syllabus, but I felt comfortable with my actions. Bottom line: There is a subjective element to the grading process, one that will sometimes put you at odds with objective notions of fairness.

9   Managing and Responding to Requests by Students Seeking to Improve Their Achievement-Related Outcomes

Sharon Nelson-Le Gall and Elaine F. Jones

We each are highly experienced instructors of undergraduate- and graduate-level psychology courses. We have held faculty positions at universities located in different regions in the United States, including the Northeast, Southeast, and Midwest. Collectively we have held faculty positions in psychology departments at large and midsize universities, where conducting research and training psychology doctoral students is a major aspect of one's professional activities, and at a small liberal arts university where teaching is the primary focus of one's duties as a faculty member.

Our parallel and shared experience regards the ethical challenge of responding to and managing requests, made by our students, which would favor and advantage them relative to their classmates enrolled in the same course. Increasingly, we each note challenges to instructing our psychology courses, which arise when students request additional coursework or an assignment to replace incomplete coursework to improve their grade. Typically these requests occur in the absence of an extenuating circumstance or documented excuse. Our concern is that faculty members encounter this sort of challenge often enough. Moreover, the manner in which faculty resolve such matters has important implications for training students who in the future will enter professions within the behavioral sciences. For example, on several occasions and across different courses, undergraduate students have asked us during the semester to assign to them additional coursework so that they may improve their grade. In such instances the other enrolled students in the course would not have received the same opportunity to improve their grade.

Case in point, one of us recalls that an undergraduate student enrolled in her course neither met with classmates to prepare a group class presentation nor presented any material during the presentation. The student tearfully requested an assignment in place of the class presentation despite not having a reasonable excuse for the instance of incomplete work, and seemed quite surprised when told by the instructor that coursework would not be assigned in lieu of the group presentation. The instructor allowed class time for students to form presentation groups, but the student who later requested an alternate assignment was absent that class session (and other sessions as well). Moreover, it was at the suggestion of the instructor (who noticed weeks later the student had yet to join a presentation group) that the student sought a group to join. The student joined a group but, despite being welcomed to work with the group, did not complete group presentation coursework. Those occurrences in the absence of reasonable excuse led the instructor to not assign coursework in lieu of the group class presentation so as not to accommodate a student's disengagement and irresponsibility.

We each recall instances in which undergraduate students have sought to have a course grade changed after the final grade roster was submitted to the registrar. While a rare and extremecase, one of us experienced behavior bordering on criminal, when an undergraduate student wanted a final grade in the course changed to a higher grade after the semester ended. The student did not dispute the validity of the recorded grade and did not want to complete additional or revised coursework. The student wanted only to receive a higher grade, believing that the submitted "B" grade would lessen the chances of admission into neuropsychology graduate programs. The student expressed the sentiment that being granted the higher grade under these circumstances was clearly warranted. The student would not take no for an answer, and the demands for a grade change escalated to stalking and harassment. In the investigation that preceded the referral of the student to the student judicial board, it came out that the student apparently had experienced some success obtaining grade improvements from instructors in the past simply by asking.

Similarly, graduate students have made requests to have "incomplete" grades for a required course changed to a passing letter grade based on work done outside of the course that had no demonstrated, direct bearing on the content of the course in which the "incomplete" grade was assigned. One of us received the request from a graduate student, who had received an incomplete grade in her course nearly a decade earlier, to be allowed to fulfill the outstanding course requirement. The student had made the request a few times in the past, but, although the request was granted, the completed assignment was never submitted. When the last request was received, the student evidently was finally intending to schedule the dissertation defense but was unable to do so because of the incomplete grade. Once again, the student never submitted to this instructor the outstanding coursework required to receive a grade for the course. Several years later it came to this instructor's attention that somehow, within weeks of the request, the student had been allowed to defend a dissertation, was awarded the PhD, and had exited the program.

We have not accommodated students who request additional coursework to improve their grade, whether during or after the semester, if other students enrolled in the same course would not have same opportunity. We believe such accommodation creates a situation of advantage for some students and disadvantage for others. Nor have we altered students' final grades upon request without justification so that it would appear they have mastered competencies and levels of achievement that were not demonstrated in our courses. We believe this type of accommodation not only creates a situation of advantages and disadvantages among the students enrolled in a course or program of study; it ultimately denies the student to whom the unearned grade is given the opportunity to experience achievement based on his or her own efforts.

We are cognizant that future behavioral scientists enroll in our undergraduate and graduate courses, and we believe it matters how we manage and respond to course-related requests that have the potential to advantage some students and disadvantage others, or to inappropriately indicate that students have been exposed to and mastered course content at a level that they have not. That (ir)responsibility matters, and the notion that achievement-related outcomes (e.g., grades) are, in part, a function of factors such as one's performance and hard work are important realizations for students to grasp as they pursue advanced training and someday enter the profession as behavioral scientists. When we instructors honor and evenly apply our course policies for all the students enrolled in our courses, then students are more likely to respect and comply with established codes of honor and ethical principles. In turn, we can hope that how we manage challenging requests and the rationale we provide when resolving issues will promote ethical behavior within the classroom and in the workplace when our former students enter the profession as behavioral scientists and become involved with conducting research and teaching students.

If in a similar situation again, we would do the same thing and not modify coursework requirements to allow a student to improve his or her grade while all other enrolled students in the course do not receive the same opportunity.

Possible general principle: Do not accommodate requests from students for additional coursework so that they may improve their grades or maximize other achievement-related outcomes while your other enrolled students do not receive the same opportunity.

10 Are There Times When Something Is of Greater Importance Than the Truth?

Bernard Weiner

The moral principle of telling the truth, particularly given scientific issues, seems incontrovertible. Here I share two illustrative ethical dilemmas I faced related to communicating the truth that call this rule into question.

Giving a "Fair" Evaluation

When I was a beginning assistant professor, a well-known athlete who also was a wonderful role model was in my class. Unfortunately, he just missed the cutoff for a grade of C. I considered his life situation, his many gifts to the school and the community, and altered the curve so that everyone at his performance score received a C. He went on to graduate (which is likely to have been the case even if I had given him a D) and became a world-famous sports icon and generous contributor to society.

Was I fair? Is it morally wrong not to give a "deserved" grade? How could I justify the vilified position of athletic favoritism? Would I have made this change if one of the other students asked for it? These questions certainly bothered me, and I attempted to maintain my moral ground with a vague utilitarian explanation that a greater good was at stake. Much would (could) have been lost if this student received the arbitrary cutoff grade.

This deservedness issue has resurfaced many times in my career but thankfully never again because of an athlete. On numerous occasions, I have agreed to pass a graduate student on the final orals when I considered the thesis and/or performance not to be of passing quality. I found myself again asking a functional question: What is to be gained (lost) by holding this person back at this time? I vividly remember facing this moral conflict in the case of a single mother, struggling with two children, who badly needed to complete school and enter the job market, yet was not performing to my standards. I agreed with the "pass" decision.

I adopt the position that, in some situations, strict adherence to the rigid tenets of an "always tell the truth" principle and the rule of deservedness is not (may not be) the best (most sensible) course to follow. Of course, this opens the door to many grey areas, arbitrary decisions, and reasonable criticisms that I (and many others) must live with. After all, it could be readily argued that these low-performing individuals are well aware of what is at stake and are expected to take personal responsibility for their actions. Nonetheless....

Withholding Scientific Data

A very different situation, also eliciting utilitarian-vs-truth concerns, regards the reporting or withholding of scientific data. Not very long ago, I collaborated on a research study with a postdoctoral student from Europe. The data were systematic, and the student was responsible for writing the initial draft of a paper intended for publication. In this draft, two of the dependent variables that failed to yield supportive or interpretable data were not included. When I brought this to his attention, he stated that these data were intentionally omitted because the study was less likely to be published if all the data were included. He further said that he had done this often in the past, with the agreement of his collaborators, and this was quite normative in his country.

I disagreed with his position and argued that the nonsignificant findings should be reported. But it also is the case that I do not apply this rule in all situations. Some research is so extensive that it is not possible to include all the findings in one publication. Further, other research is so expensive that some data are gathered because of the opportunity and are "saved" for later inquiries. The question I asked the student was: Is it better for the growth of science, better for the understanding of the theory being tested, if these data were reported? This is again a utilitarian question, but in this case the well-being of the science rather than the agent is the focus of concern (although both situations involve the total welfare of society).

I believe it is possible that, even given a small study, the betterment of science may not be served by including all the data if these data detract from the focus of the work or create more noise than light. Often I have read research articles with so much peripheral data that the main message is lost. For this reason, I ask my students not to collect data if they are not highly pertinent to their hypotheses. Of course, often this is an inefficient and costly way to proceed. But it prevents the dilemma that my postdoctoral student faced, where his intention was not only to build science but also to build his publication record.

How was the conflict with the student resolved? I could not (did not) "demand" that he refrain from publishing the study as is, or that he "weaken" his hopes for publication by including data he wanted to omit. I explained my position and withdrew my name as a coauthor, hoping that my actions might alter his position or at least convey my strong beliefs about the priority of scientific progress versus personal advancement of the scientist. In this case, the moral course was full conveyance of the truth because these nonsignificant data added to scientific understanding. Of course, it is easier for someone who has a job, has tenure, and is a full professor to take this position than for someone hoping to find a place in the world. In sum, I believe it is not unethical to withhold some data in some research contexts if the intent is to further the progress of science, although this is a difficult judgment for a potential author, fraught with hedonic biasing.

To complete the story, the paper was accepted without these data, and the student has gone on to a successful career. Is he still withholding data? I do not know, although I must admit to some hesitations when reading his work. Is there something else I could have done, something more coercive, to alter his actions? Perhaps, but I did not want to destroy our relationship, nor did I want to be perceived as a career obstacle, and I do not know what an alternative course of action might have been.

11 Commentary to Part II

Susan T. Fiske

"Fairness to all" is an agreed-on concept in the abstract but much more fraught with ambiguity to apply in concrete instances. This section reaches some consensus on using the fairness doctrine to resist students who plead for special treatment, even in the uncomfortable decision to require that they document their medical misfortune, and certainly in the seemingly callous refusal to allow making up for past underperformance due to lack of effort. Ad hoc bargains with individual students simply are unfair to the rest of the class.

In the same way, secondhand information as evidence in ethical decision making requires documentation and confirmation, again in fairness to all. Just as some people are manipulators, others are saboteurs. The collective needs to be protected from those who try to game the system, and verification is key here.

What seems right in the abstract, however, is complicated in practice. In regard to grading, this section raises dilemmas of how to credit extraordinary improvement or effort. What about humane adjustments to a borderline grade, considering exceptional life circumstances? If an instructor does bend the rules, what does that teach the students, not only about academic ethical standards but also about ethics in their own later careers?

Applying the fairness doctrine in an even-handed but humane way may indeed create a teachable moment for all concerned. Beside fairness, all parties need to have the same implicit or explicit contract to guide their choices. Playing by the same rules requires knowing the rules in the first place.

Part III

Authorship and Credit

12 An Ethical Dilemma in Publishing

Larry E. Beutler

As a newly appointed associate professor in a well-known medical school's department of psychiatry, I was assigned by the chair of the department to a prominent research laboratory. This position required that I learn a new body of research and participate in the ongoing research programs of the laboratory as well as developing my own funded and unfunded research programs. One day, during a conference on one of the patients seen in our laboratory, the laboratory director, a very prominent psychiatrist, asked if I would be willing to review a book for a prominent medical journal. This was a very prestigious journal, and the book was on a topic that I was beginning to research, so I believed I knew the literature reasonably well. It also occurred to me that publishing this review would enhance my visibility and reputation in the psychiatric community. Based on these perceptions, I was eager to immerse myself in the process; I agreed and gratefully took the rather large volume and spent the next two weeks reading the book and articulating a rather thorough commentary on its strengths and weaknesses.

I was very pleased when, two days after I turned the review in to my laboratory director, I received his thanks and compliments for the thorough job I had done. He indicated that he would forward the review to the journal within the next few days. Nothing more was said about the matter until I came across my review while reading another article in the medical journal to which the review had been submitted. Lo and behold, there was my carefully crafted review of the book – not a word had been changed, except.... I was not the author. The laboratory director's name appeared prominently where mine should have been.

This flew in the face of what I conceptualized as ethical behavior. I immediately talked to the chief of psychology in the department and expressed to him my concern with having my work misrepresented. He recommended that I let the matter drop and learn a lesson from the process. His fear was that discord between the disciplines of psychology and psychiatry would erupt and would not be good for the psychology division. He explained that medical laboratories frequently operated according to different ideas of ethics than those in PhD programs, and that it was not unusual for a laboratory director to have his name listed first on all publications coming from his laboratory. I indicated that I understood that possibility, but that this was a case of his having claimed *sole* ownership to my work, a condition that was clearly in violation of the American Psychological Association's code of ethics, and I assumed of the American Psychiatric Association as well.

Finding no solace in the comments of the chief of psychology, I took the matter up directly with the laboratory director. He was taken aback by my unhappiness and virtually scoffed at my "narcissistic injury," asserting that it was his right and place as director of the laboratory to be listed as the first or only author on all papers that came from the lab. And because there was only a place for one reviewer, it was clear to him that the review should carry his name, representing the laboratory. Although I continued to work in the laboratory and published many papers on the research projects that were conducted there, the incident forever strained the previously warm relationship that I had with the director. I learned some important lessons from this event, however, and took them to heart through the rest of my career.

First and foremost among these lessons was to *never assume*. I had unjustifiably assumed that all professionals accepted and followed the same basic ethical standards as did I. I had also assumed that the implicit contract into which I entered in taking on the task of reviewing the book was understood by both my director and me. Both assumptions, as it turned out, were false. While I followed the common practice asserted by the ethics of the American Psychological Association that the order of authorship should reflect the effort expended, I came to understand that it is not an uncommon practice outside of American psychology for the laboratory head to be the designated first author of all papers that are published, whether or not he or she participated directly in the task. This practice is not ordinarily considered unethical in circles outside of psychology.

I also learned that it is important at the outset of a project to establish the requirements for authorship before a task is initiated. On many subsequent occasions, I have found myself in a position with my students where commissioned articles are to be written and they are given an assignment to write the first draft. On most occasions, my rule of thumb has been that if they write the final draft, they should be the first author, but there have been two exceptions to this rule, in my recollection. On one occasion, I wrote the penultimate draft and left it to be finished by the second author. This decision was occasioned by the explicit request by the editor who invited the submission that I maintain my role as first author. I have since adopted the general stance, based on several similar requests, that when the manuscript is for an invited paper that is specifically designated to represent my contributions to a collective area of knowledge in the field, or if the paper entails the primary report of a specific empirical research program on which I am the principal investigator, I take responsibility for preparing the final manuscript. I believe that it is also imperative that, as the laboratory director, I personally oversee, review, and contribute to all other steps in the process of analyses and preparation. The assignment of authorship beyond this point is, as is conventional in the field, done on the basis of effort expended. I reserve the right to assign authorship to students based partly on who does the first draft and partly on the workload assumed in the process of gathering an analyzing data. While I continue to be a bit lax in following my intention of defining in advance the order of authorships on various papers, the few times that this has been a problem have been easily resolved by making sure that all students working on projects in my laboratories know the working principles that underlie decisions about authorship, as articulated above.

13 What Does Authorship Mean?

Dale C. Farran

The issue of what authorship means – who deserves credit and what one's name "hanging from the masthead" (Burman, 1982) actually means about contribution to the article – never came up in my graduate program. Since the 1980s there has been a continuing spate of articles about the ethics of authorship and concern about the increasing number of authors listed on individual papers (Bennett & Taylor, 2003; Holaday & Yost, 1995; Marušić, Bošnjak, Jerončić, 2011; Smith, 1994). Most of these concerns have been expressed in the medical or life sciences, less often in the social sciences, and all of them involve unwarranted credit for authorship – authorship as a "gift." The ethical issue about authorship I am presenting is different. It concerns whether the content of the article is something all authors agree with, which, I assumed, one's name as an author would mean (Janssens, 2014).

As a new relatively inexperienced PhD graduate, I was given the opportunity to supervise the continued collection of data in a longitudinal study. My portion, of course, was small and not part of the primary data being collected on the sample. Nevertheless, it seemed as though I could not only make a contribution but also achieve some recognition for it in the way of publications. I was working with a strong senior scientist in the field, someone with whom it would be good to link my name in journal articles.

Although the original sample was homogeneous, we collected data on an additional group composed of individuals who were quite different from those in the primary one. We used the same measures and, as might be expected, the two groups appeared to be very different from each other. We had collected two kinds of data. One set of measures involved interactions that were presumed to be predictive of scores on the second set of measures – developmental outcomes. Although I had collected some of the data, my mentor had conducted the analyses and drafted the results section of a paper to be submitted to a journal.

When I read the draft, I realized that we had treated as a single group a heterogeneous sample composed of two groups who differed on both the presumed predictors and the outcomes. Not only that but the group variable had not been included in the analysis. When analyzed as a total group, there appeared to be a strong relationship between the interactive behaviors and the outcome. Moreover, there was a certain face validity to the findings – the relationship was believable, it made sense in that it fit most people's worldview.

However, if the relationship was a true one, I argued, then it should operate similarly in each sample if the groups were examined separately. There was a problem of sample size; each of the samples was relatively small, but this should only affect the significance levels. If the connection between the two set of variables was as strong as it appeared, then the correlation should be of similar magnitude within each of the samples. Within each sample, there was variability in both the predictive variables and the outcomes (though not much overlap because of the mean differences in each). When the samples were examined separately, however, the correlations were essentially zero; within each sample, there was no relationship between the predictor behaviors and the outcomes.

I contended that we could not submit the article with the two samples combined. Though it may seem strange to consider that the article could have been published with this substantial flaw, it is important to note that this incident occurred a long while ago. Social science research was far less sophisticated, and quite a bit of presumed knowledge was derived from just these sorts of analyses of heterogeneous samples. My arguments were met with resistance – the relationship just made so much sense! After wrestling with the issue and with genuine fear that I was placing my career in danger, I announced that if the article moved ahead it would have to do so without me as a coauthor. I removed my name from the manuscript.

This step may seem small and inconsequential, but I assure you it was not. I was at the beginning of my career. The person with whom I disagreed was an established figure, well published, and influential. My decision questioned my collaborator's behavior and credibility. In the long run, one could argue that it did not really matter. Here was a publication possibility. The press to publish was beginning to take hold much more strongly in the social sciences. Was it really important that the conclusion of the paper was inaccurate and misleading? I would have another entry to my vita. The half-life of all but a tiny few publications is much less than even a few months. Many articles are read and cited by almost no one else in the field. This incident occurred in the era of counting vita entries and before the era of examining citation rates.

The ethical dilemma for me was that I really believe in the scientific enterprise. I believed then and still do that individuals contribute small bits to the blocks that build the edifice of knowledge. We cannot always be correct, and often what we thought was a contribution is later rejected, but it is essential initially as social *scientists* that we believe in the truth of what we write. If we know at the outset that something is likely untrue, it is unethical to present it, even if the field is highly receptive to receiving it. Maybe, in fact, in just such times of high Zeitgeist credibility, we have to be the most vigilant about our contributions.

In terms of my own ethical dilemma, the paper was not published as originally written. Instead, a year or so later, the manuscript was reworked examining the processes within each sample and a different conclusion was reached, perhaps not as strongly as I would have liked but one I could live with. I was a coauthor of this paper.

But I still feel trepidations about my decision. As I read various discussions of authorship – "authors" finding their names on papers they have never read (Janssens, 2014) – I realize how fragile our science is and how important it is to have a strong and well-centered set of beliefs about what your name means when it is on a piece of research.

REFERENCES

Bennett, D., & Taylor, D. (2003). Unethical practices in authorship of scientific papers. *Emergency Medicine*, 15, 263–270.

Burman, K. D. (1982). Hanging from the masthead – reflections on authorship. *Annals of Internal Medicine*, 97, 602–605.

Holaday, M., & Yost, T. (1995). A preliminary investigation of ethical problems in publication and research. *Journal of Social Behavior and Personality*, 10, 281–291.

Janssens, C. (2014). Let's clarify authorship on scientific papers. *The Chronicle of Higher Education*. Retrieved April 11, 2014, from http://chronicle.com/article/Lets-Clarify-Authorship-on/148287/?cid=wb&utm_source=wb&utm_medium=en

Marušić, A., Bošnjak, L., & Jerončić, A. (2011). A systematic review of research on the meaning, ethics and practices of authorship across scholarly disciplines. *PLoS ONE*, 6(9): e23477. doi:10.1371/journal.pone.0023477

Smith, J. (1994). Gift authorship: A poisoned chalice. *British Medical Journal*, 309, 1456–1457.

14   The Ethical Use of Published Scales

Diane F. Halpern

It often seems that trouble finds me when I am not paying attention. Like many faculty, I am often asked to chair a wide variety of undergraduate theses – often on topics in which I have very little expertise. Several years ago, a bright student wanted to explore the variables that influence assimilation in a group of immigrants that had been largely ignored by psychologists, probably because the group is not large. She did a terrific job getting a subject pool together, especially given that it meant ferreting out participants from around the country. In her first study, she used a scale that was widely used and appeared in many publications; its popularity was probably because the scale had reasonable psychometric properties and there were few alternatives beside writing one's own scale and then testing its properties – which can be a study in its own right. Encouraged by success in her first study, the undergraduate conducted a follow-up study, this time modifying the scale slightly so that it was more directly related to the specific life experiences of this understudied group. I believed that although the results were not groundbreaking, they would be interesting to scholars in assimilation and cross-cultural adaptation and would extend the psychological knowledge about this small group.

The outstanding student researcher was a senior at the time and wanted to apply to selective graduate programs. Thus, it was important for her personal success that she publish her study. As most readers probably know, graduate admissions committees value publications because they provide evidence that the student applicant can conduct solid research. Of course, the student was first author because it was her idea, and she did all of the real work. I commented on various drafts and made sure that the study conformed to all of the usual ethical requirements of psychological research, including approval from the Institutional Review Board and informed consent from participants.

A write-up of the research was sent to an appropriate journal with all of the fanfare and trepidation that accompanies someone's first attempt at publication. (I am referring to the undergraduate here – I had a reasonable number of publications at the time, which means that I should have seen trouble coming.) As seasoned psychologists probably expect by now, the manuscript was sent to the author of the scale we used. Of course, the author of the scale was consistently acknowledged everywhere the scale was mentioned, and we included several references to articles that had used the scale. I was surprised by the angry review by the author of the scale, who claimed that I had behaved unethically by using a scale that was written by someone else without obtaining permission, by revealing the items on the scale in the proposed publication, and, in the second study, by altering the scale without the author's permission.

Was any of this unethical as the reviewer/scale author claimed? Actually I was not sure. All of the scale items had appeared previously in other journal articles. I always thought it was a good idea to include scale items in a manuscript so readers could see what was asked and how the scales were constructed. Did I need to obtain permission from the author of the scale before I could use it? We carefully noted where we made changes in our second study, but did we need to do more, or maybe less? I was confused by the allegation that I had behaved unethically, and the poor undergraduate was sure that her career was doomed – no graduate school would accept a student who had behaved unethically.

I hope that readers are pondering the ethical questions raised by the reviewer, who threatened that he would complain to the Ethics Committee at the American Psychological Association. Fortunately I knew several experts who could help me figure this out. It is hard to keep the experts' identities anonymous because of the positions they occupy, but their identity is not a problem because they were never accused of unethical conduct. The head of publications at APA assured me that I had not violated any ethical guidelines, not in the letter or even in the spirit of the guidelines, because all of the scale items had been published previously. I also contacted the director of the Science Directorate at APA, who independently gave me the same assurances. These opinions were sent to the irate reviewer, who never raised the issue again.

But what about the manuscript? I might have been happy to allow it to languish in my "drawer of damned manuscripts" (actually an electronic file), also known as "studies I have conducted but never published." Maybe I will brush up some of the better ones in my old age so they can see the light of day, or maybe I will just let sleeping (and usually flawed in some way) studies die a natural death. But this manuscript was not solely mine; there was an anxious student who had hoped that it would be a ticket to graduate school and a stellar career in psychology. Ultimately, it was published in another journal, and we never heard from the infuriated reviewer again.

You may be wondering about the student who found herself in this maelstrom. I am happy to report that her career is soaring. She just may be teaching at a university near you.

15 Idea Poaching Behind the Veil of Blind Peer Review

Rick H. Hoyle

Productive researchers are called on with regularity to participate in the peer-review process for academic journals. Generally, journal editors invite feedback on manuscripts submitted for publication from scholars whose research interests align with the focus of the new research or theoretical model presented in manuscript. In some cases, either by journal policy or author request, the review process is fully blind. That is, the authors of the manuscript do not know who reviewed it and the reviewers do not know who wrote the manuscript. Typically, however, the process is only partially blind, with researchers allowed to see the names of the authors of manuscripts they are invited to review. The privileged access to new, potentially important ideas in one's area of research without the awareness of the originators of those ideas may tempt some researchers to share the content of the manuscript with colleagues or even make use of the ideas in their own work without permission from or credit to the source. This temptation typically is held in check by ethical guidelines of professional societies and journal publishers that govern the handling of manuscripts by reviewers. On occasion, temptation overcomes ethics, and researchers abuse reviewer privilege. Early in my professional career, I was a naïve witness to an abuse in the form of idea poaching.

I was collaborating with a senior scientist doing research on a topic with which he was strongly identified. Most of his research at the time was on this topic. Given his well-earned reputation as an authority on the topic, he was often called on to review related manuscripts. To one of our regular meetings he brought two copies of a manuscript he was reviewing for a top journal in the field and suggested that we look it over together. At that time, I had little experience with the manuscript review process, and therefore did not question the appropriateness of looking through a manuscript that he had been asked to review. In fact, I felt privileged and trusted for having been allowed to see the document.

After a lengthy discussion about the authors' perspective, approach, and findings, our discussion settled on a specific idea the authors proposed in their framing of the manuscript. But rather than focusing our discussion on the merits of the idea, my senior colleague asked how we might incorporate the idea into our own thinking on the topic. In fact, he suggested, it might be possible to make adjustments to our ongoing studies in order to provide a more compelling test of the idea than those offered by the authors in this manuscript. I recall having strong reservations about this suggestion. At a later meeting I expressed to him my reservations, arguing that we would be using someone else's idea as if it were our own. He responded that there is no ownership of ideas and that what really mattered was what was done with the idea. In his view, the authors had taken an idea available to anyone but had not made good use of it. As they did not own the idea, it was our prerogative to put it to better use.

Two ethical guidelines regarding the handling of manuscripts by reviewers were violated in this situation. The first, sharing of a privileged document, was not evident to me at the time. As a result, I joined my senior colleague in an invasion of privacy authors assume they are guaranteed when they allow a journal to consider their manuscript for publication. The second, using someone else's idea without their permission after having learned of the idea through access to a privileged document, was evident to me even without formal knowledge of professional or journal ethical guidelines governing the manuscript review process. Although I made evident my concern to my collaborator, when he justified the action, I did not pursue it further. My reactions, or lack thereof, can be attributed to naiveté and on the first count and hesitancy to assert my view from a position of low power on the second.

Having now reviewed many manuscripts and served as editor for two journals, I am well aware of the responsibilities that come with the privilege of reviewing manuscripts submitted for publication. In Standard 8.15, the Ethical Principles of Psychologists and Code of Conduct (APA, n.d.) addresses both ethical issues raised in the situation I described, stating that "Psychologists who review material submitted for presentation, publication, grant or research proposal review respect the confidentiality of and the proprietary rights in such information of those who submitted it." Similarly, in its Ethical Guidelines for Reviewers, the American Association for the Advancement of Science states:

> The submitted manuscript is a privileged communication and must be treated as a confidential document. Please destroy all copies of the manuscript after review. Please do not share the manuscript with any colleagues without the explicit permission of the editor. Reviewers should not make personal or professional use of the data or interpretations before publication without the authors' specific permission (unless you are writing an editorial or commentary to accompany the article). (AAAS, n.d.)

Although these statements make clear the lines should not be crossed when handling manuscripts submitted for review, they offer little in the way of practical advice for behaving ethically when tempted to share the contents of a manuscript in review or use ideas gleaned from those manuscripts. One might reason that the ethical high road is to contact the authors directly about their ideas or data; however, such contact circumvents the blind review model promised by most journals. Instead, Rockwell (n.d.) suggests contacting the journal editor and asking him or her to approach the authors about their willingness to interact with someone who has served as a reviewer of their manuscript. In their Ethical Guidelines for Peer Reviewers, the Committee on Publication Ethics (COPE) echoes this advice: "Peer reviewers should not involve anyone else in the review of a manuscript, including junior researchers they are mentoring, without first obtaining permission from the journal" (Hames, 2013). In terms of idea poaching, Rockwell asserts, "You cannot use the information in the paper in your own research or cite it in your own publications. This can raise serious ethical issues if the work provides insights or data that could benefit your own thinking and studies." Neither source offers advice for managing these violations when asked to participate in them by someone else; however, it seems reasonable to infer that, minimally, one should recognize that such behavior violates ethical standards and refuse to participate in it.

Because I am now a senior scholar, it is hard to say with certainty what I might do if I could relive that situation at that point in my professional development. I would hope that, given the increased attention to ethical practice in psychological science, a faculty mentor or perhaps graduate coursework would have informed me that manuscripts under consideration for publication could not be shared with others. Were that the case, I would have been aware of both violations of reviewer ethics in which I became complicit. But, given this awareness, what could I do, given the power differential between a senior scholar, who was paying my graduate stipend, and a fledgling? I could have asked my collaborator whether he had permission to share the contents of the manuscript. If he confessed that permission had not been given, then I could offer to review it with him but only if the journal editor would allow it. (It is now common practice by many senior scholars to ask for such permission and conduct reviews collaboratively with senior graduate students and postdocs for the purpose of giving trainees supervised firsthand experience reviewing manuscripts.) Whether or not my senior collaborator had been given permission to involve me in the review of the manuscript in question, the ethical violation concerning the use of the authors' idea without their permission and as if it were our own would have remained. As I did at the time, I would voice concern about this behavior. If expression of my concern did not prompt appropriate action, I would suggest that we ask the journal editor to contact the authors and inquire about their willingness to discuss their idea with interested colleagues or perhaps even collaborate. If that permission were either not requested or not given on request, then I would indicate an unwillingness to pursue new research based on use of the idea. Although it is not clear that I should go further and report my collaborator's misconduct to the journal editor, it seems unlikely that I would do so given the potential fallout.

Publication in academic journals is the primary means by which scholars contribute to the knowledge base in their discipline. The quality of published work is maintained by the review of papers submitted for publication by knowledgeable peers. Because the interests and expertise of peer reviewers is often in the area covered by manuscripts they are asked to review and, importantly, because their identity generally is not known to authors of the manuscripts they review, there are temptations to share the contents of manuscripts submitted for publication with collaborators and poach ideas therein. In recognition of these temptations, professional organizations and journal publishers have established ethical guidelines that serve as a deterrent to reviewer misconduct. Because of the collaborative approach to research characteristic of the social and behavioral sciences, early-career scholars might be asked by senior collaborators to violate these guidelines. Knowing the guidelines and refusing to participate in violations of them are critical to maintaining the fairness and trust necessary for effective peer review and appropriate attributions for new ideas.

REFERENCES

American Association for the Advancement of Science (n.d.). *AAAS ethical guidelines for reviewers.* Retrieved June 27, 2013, from http://www.sciencemag.org/site/feature/contribinfo/review.xhtml

American Psychological Association (n.d.). *Ethical principles of psychologists and code of conduct.* Retrieved June 27, 2013, from http://www.apa.org/ethics/code/index.aspx?item=1

Hames, I. (2013). *COPE ethical guidelines for peer reviewers.* Retrieved June 27, 2013, from http://publicationethics.org/files/Ethical_guidelines_for_peer_reviewers_0.pdf

Rockwell, S. (n.d.). *Ethics of peer review: A guide for manuscript reviewers.* Retrieved June 27, 2013, from http://ori.dhhs.gov/sites/default/files/prethics.pdf

16 An Ethical Challenge

Susan Kemper

The headlines in the *Chronicle of Higher Education* sometimes scream "MY ADVISOR STOLE MY THESIS," but I generally dismissed these claims, assuming the advisor had made substantial contributions to the thesis, which were overlooked by the student. Then a student collapsed in my office, sobbing, "My advisor stole my thesis." She added, "And submitted it to [a federal funding agency] ... and then hired me to do something else." I had served on the student's Masters committee and knew her history: the advisor had been on sabbatical leave during the year that the student had conceived, executed, and written up the thesis; I knew that the student had complained that the advisor had not responded to e-mail messages, had provided little guidance, and had not reviewed drafts in a timely manner. I had been urging the student for some time to switch advisors, but she stuck with him because he was a recognized authority in the field. Now this: the advisor had submitted the thesis to a federal agency as an application for a small grant; when it was subsequently awarded, he hired the student and assigned her a new line of research. She protested, arguing that the grant was awarded for work – her work – already done and the new line of research was not specified in the grant application. The advisor dismissed her concerns telling her, "This is how it works. Everybody does it."

This situation raised a number of concerns: how to protect the student from retribution yet ensure she received appropriate credit for her thesis; how to inform the funding agency; what to do about my colleague. At my urging, the student compiled a timeline, including a trail of e-mail messages to the advisor that had been ignored, and electronic drafts of the thesis bearing timestamps as well as a copy of the grant application. We then met with the vice-chancellor for research, who appointed a committee to investigate the student's allegations. Meanwhile, the student was shifted to a teaching assistantship. And the grant – which after all was awarded to the university – was declined. Neither I nor the student ever saw the committee report, nor were we informed of the outcome of the investigation. But by the end of the academic year, the advisor had left the university, taking a similar position at another. And, unfortunately, the student also left the university, abandoning her research career to become a high school teacher.

If I found myself in this situation again, I would insist that committee findings be published and the advisor be publicly sanctioned if the committee concluded there was wrongdoing. From my perspective, our problem merely became another university's problem. I have always felt I failed the student; her thesis was never published and she left the university believing "everybody does it."

Possible general principle: Misconduct in science is a problem that cannot be ignored and overlooked.

17 Authorship: Credit Where Credit Is Due

Stephen M. Kosslyn

Some years ago I was asked to mediate a heated dispute between a postdoc and a graduate student in my lab. Both agreed that they should be authors of a report of an experiment, but they disagreed about who should be first author. Resolving this dispute led me to devise a set of criteria for determining who should be an author and the order of the authors. I posted these criteria on our website, which made all the difference. Going forward, in the few instances where contributors disagreed, simply discussing these criteria invariably resolved the issue.

In my lab, we consider six specific criteria and assign points to each; often the points for each phase are divided among several people. The point totals for each phase should be discussed – and possibly adjusted – as part of this process. For example, some projects use standard designs (e.g., a "Stroop" task) or analyses (e.g., correlations), in which case the number of points for that phase should be reduced.

The following are "default" point values, with a total of 1,000. Points are assigned based on creative contributions to a specific phase. Points for each phase are divided among authors in proportion to their creative contribution to that phase of the project. The ordering of authorship is determined by the relative number of points.

1. The idea (250 points): Without the idea, nothing else happens. If the idea grew out of a discussion, all who contributed get "credit" – but perhaps not equally so.
2. The design (100 points): The details of the design include counterbalancing issues, control conditions, whether a within-participants or between-participants design is used, and so on. A bad design later will render the results useless, so this is a critical step.
3. The implementation (100 points): Someone must implement the design into actual materials, devise instructions, and so on. To the extent this is simple boilerplate (a variation on well-developed methods using available materials), this step may be given much less weight. The person who implements the design may be supervised

closely by someone else; if so, some of the points may go to the supervisor.
4. Conducting the experiment (100 points): The person who tests participants *can* earn up to 100 points, but may not earn any points if all he or she does is mindlessly carry out instructions. Authorship is awarded only to those who contribute substantially and creatively to a project; if someone is receiving class credit or payment and all they do is follow instructions and test participants, this is worthy of an acknowledgment in the footnote, but not authorship. On the other hand, if this person notices what participants are actually doing and offers an original interpretation for the results, diagnoses a problem with the design or procedure, makes useful, constructive suggestions about how to repair a problem, observes interesting hints about what's really going on in the debriefings, and so on, this would count as a substantial creative contribution.
5. Data analysis (200 points): Simply running the data through an ANOVA program is not enough to earn authorship at this phase. However, devising a new way to look at the data (e.g., as difference scores or ratios of some kind) or otherwise contributing a novel insight into the best way to reveal the underlying patterns in the data may be sufficient. Particularly labor-intensive or creative data analysis, such as those that are sometimes required in neuroimaging studies, can earn the full number of points. Depending on the project, the maximum of 200 points may or may not be allocated.
6. Writing (250 points). The research will have little or no impact if the results are not formally reported. Writing is usually shared by several people. Credit is allocated primarily to the one who shapes the conceptual content, although a good and insightful literature review also counts heavily. If someone writes a first draft that is not used at all, this does not contribute toward points: good intentions are not enough. Similarly, the sheer amount of time one has spent on the project is not relevant; competent people who work more efficiently should not be penalized.

To qualify as an author of a research report, a person has to make a creative contribution in one or more of these key phases of research; a lesser contribution warrants an acknowledgment in the footnote of the paper. The lab head determines whether someone deserves either of these credits, and determines the ordering of authors according to their contribution to each of the key phases. As noted earlier, I assign a larger weight to the first and last phases, and to any other phase that requires special expertise or creativity (e.g., data analysis, in some cases). The key notion is replaceability: If someone else could have done the job just as easily, that contribution is weighted less highly than a contribution that hinged on special insights, talents, or abilities.

The key to fair allocation of authorship and equitable ordering is to have criteria that are known to all and that all can discuss. In addition, when the ordering is not obvious (but it usually is), each contributor can send the lab head an assessment of his or her contribution after the project is relatively complete but before the paper is written. If it's not obvious what the ordering of authors will be, the contributors can discuss this and discuss giving one of them the opportunity to take a larger role in the writing or data analysis process – thereby allowing him or her to accrue more points.

In short, with transparency and a little thought, authorship disputes can be avoided and everyone involved can feel that the process of allocating credit is fair and principled.

18 Publication of Student Data When the Student Cannot Be Contacted

Peter F. Lovibond

A colleague of mine once lost a graduate student. No goodbye note, no forwarding address, no response to repeated e-mail messages, phone calls, or letters. Not even the power of Google could locate this student. So what was my colleague to do with his student's project? Could he publish it? Could he correctly interpret the results or vouch for their validity? He didn't even know for sure if he had the final version of all the data. In the end he did nothing. So all of the work that the student put into the project, and all the work that my colleague put in, was effectively wasted.

Since that episode, I have come across several similar but less dramatic examples, and I have also experienced something similar myself. The most common scenario is that a student remains in touch with the supervisor for a period of time after graduation but fails to write up his or her data. Then at some point the supervisor discovers that the student can no longer be contacted. The ethical dilemma of course is whether the supervisor can take over and write up the student's data for publication. On the one hand, it seems unethical for the supervisor to share credit for work done by another individual, and in particular to proceed with publication without the input and approval of that individual. On the other hand, it seems unethical to allow good research, often funded by the taxpayer, to remain unpublished. Not only that, but the supervisor has usually made a substantial contribution to the project that cannot be rewarded through publication.

Recently I decided to raise this issue in my university. The response I received was that the university's research code stipulates (1) that no person who meets the criteria for authorship may be excluded without their permission, and (2) that all authors must approve the final published version of the paper. In addition, at my university, students own the data they collect so nobody else can use their data without their consent. Many journals have similar requirements. So I had a clear answer: If you lose contact with a student, you cannot submit his/her results for publication either with or without the student being named as an author.

Interestingly, most academics with whom I have discussed this issue disagree. They think that the supervisor should have the right to publish a student's data if the student has not done so within a reasonable period of time. They also agree that the student should be named as an author, with their position in the authorship list reflecting their overall contribution to the paper – in other words, not necessarily as lead author since they would not have written the paper.

Can this dilemma be resolved or prevented? I am currently exploring this question at my own university. So far, the most promising avenue appears to be to ask students and their supervisors to sign a document concerning publication of data before embarking on the project – a sort of academic prenuptial agreement! The idea would be that students have the right to write up their own data for publication, usually as first author if they provided all of the data. However, if a student did not take this opportunity within a defined period of time, say two years from graduation, then the supervisor would have the option to write up the work, with order of authorship determined according to the normal criteria. Another opportunity for an agreement to be signed would be at the time of the student's departure, covering any material that hadn't yet been published. It would be desirable to have an independent research committee provide oversight of such a process, for example to verify that the student could not be contacted and that the work had not already been published by the student.

I should point out that it is by no means clear if this strategy would actually work – whether it would be accepted by the university, the journal, or the funding agency, if one is involved. However, I hope this brief account will stimulate discussion and perhaps lead to an accepted approach to minimize supervisor frustration and maximize the chances that good student research will be published.

19   Ethics in Research: Interactions between Junior and Senior Scientists

Greta B. Raglan, Jay Schulkin, and Anonymous

A moral education for Mead, and certainly for Dewey (1975), refers to a developmental trajectory, bootstrapping on empirical findings about development and attention. These findings need to be anchored to moral development. Education nurtures many cephalic capabilities, not least of which are our ethical sensibilities. But a moral education, as Aristotle (1999) and the Stoics noted and Dewey reinforced, necessitated the development of character, specifically a moral character in which self-corrective processes are tied to humility, where a nurtured self without a bloated head is a normative goal.

In discussing dynamics of power in research relationships, the focus is often on the roles of researchers and participants, and the importance of avoiding coercion or forced participation. This power relationship is also extremely important to keep in mind when considering the role that an advisor plays in the life and career of a junior researcher. While an advisor may use his/her power to guide, support, and educate, this also makes it very easy for an advisor to take advantage of an advisee. The examples given by an advisor or senior researcher can shape a beginning researcher's moral education and can affect the ethics of the field far down the line. In this chapter we, along with a colleague who wished to remain anonymous, provide examples of ethical lapses we have seen in training relationships.

As a graduate student, for example, a fellow student came to me to discuss an ethical concern he was having with his research advisor. The student had completed all of his degree requirements and had recently completed his dissertation defense. He had finished the revisions of his dissertation and had the signatures of all of his committee members. His advisor, in an apparent effort to increase her publication output, informed the student that she would not sign off on his dissertation until he had completed and submitted a publication draft of his findings with her as the second author. The student had had a similar experience when he completed his master's thesis project, and while he had heard that this was common practice in his lab, he was concerned because his advisor intended to delay his graduation until he completed the draft. My colleague felt, and I agreed, that his advisor was unfairly holding up his academic progress to ensure personal gain. In addition, he felt unduly obligated to assign his advisor a more prominent authorship position than he believed was warranted based on her involvement in the project. At the time, I advised my colleague to seek assistance from the head of the department to ensure that he graduated on time. The student had considered this option, but was concerned that such action could lead to retribution from his advisor, such as refusal of letters of reference, that would negatively affect his future goals. We discussed the pros and cons of confronting the issue with another faculty member; however, his concerns about possible anger from his advisor led him to complete the manuscript without notifying other faculty members of his advisor's policy. My inclination was to report the behavior to the advisor's superiors myself but, lacking verifiable support for my statements and also fearing retribution against my colleague, I kept mum. To my knowledge, his advisor was never censured for her actions, and she continued to make the same requirements for subsequent students.

With further reflection on this event, we continue to believe that the advisor took advantage of her position of authority and seemingly she believed she was beyond reproach for such behavior. Looking back, we would likely work with the student to report the violations through the appropriate channels in an effort to reduce the possibility that this might happen again to another student. In addition, I would attempt to work with faculty to ensure that there were no negative ramifications directed at the student. In spite of the advancement of time, however, I empathize with my younger self and the plight of my colleague.

A second example, reported by another colleague, is less clear-cut. My colleague was working with a large neuroimaging data set and had entrusted a junior faculty member to complete some very sophisticated image analyses. Neuroimaging data, including structural and functional MRI (as well as data from many other types of imaging techniques), are often analyzed using sophisticated data pipelines. It is often difficult to check the results at intervals along the pipeline, and there are few "gold standards" that can be applied to know for certain whether the results are valid. Sophisticated statistical methods are required, many of which involve assumptions that are often unchecked. My colleague, who has extensive expertise in statistics but not in neuroimaging analysis, questioned her younger associate about some of the analysis methods and results. The junior faculty member was defensive and unwilling or unable to provide sufficient explanation of the veracity of her methods, and went on to complete the final draft of the manuscript. My colleague, the principal investigator of the project, then asked several experts in the field of neuroimaging analysis to review the data analysis and the results. They concurred that there were errors in the method, at which point my colleague insisted that the manuscript could not be submitted. The junior member of the team maintained her right to submit the manuscript with or without the approval of the project PI. The manuscript was submitted without the PI's name on the author list. Fortunately, it was not accepted for publication. In this case, the PI did not herself have the expertise to evaluate with certainty the validity of the data analysis but had enough knowledge to call for assistance. It is essential in neuroimaging projects, which usually involve a multidisciplinary team of researchers with varied levels of expertise, for investigators to be able to trust each other's abilities and integrity. Individual researchers, even the PI, often do not have sufficient expertise to truly understand all of the intricacies surrounding data acquisition, image analysis, and data analysis. When there are serious concerns, as in this example, the PI clearly did the right thing by asking for input from experts outside of the project. It is not clear, however, what rights a PI has to refuse to allow a junior colleague to publish questionable data. In this example, however, it is clear that my colleague's position of authority contributed to her ability to confront an unethical research practice. Again, the relationship between a supervisory researcher and a less experienced colleague is an area in which ethical oversights or missteps can be challenging and are frequently faced.

The expectation is that an advisor will refrain from satisfying immediate self-concerned desires and opt for socially enhancing actions instead. In this way, the expectation is that the individual can increase his or her own reputation and become more likely to be supported by others without infringing on the well-being of others. Unfortunately, not all members of the scientific community adhere to this expectation. As a graduate student, junior researcher, or volunteer researcher, it is easy to feel that the word of one's advisor is truth, and that confrontation or disagreement is dangerous or maladaptive.

Truth telling, even under the best of conditions, is a faint motive that competes with other motivations such as academic or professional success. Several famous incidents have shown that events affecting a researcher also color the experience of his or her advisees and colleagues. Other reports exist of academic advisors encouraging students to manufacture or manipulate data, even at top institutions. The threat of persecution or retribution (either from an advisor, a senior investigator, or the scientific community) may be enough to prevent many young researchers from taking an effective stand against unethical research behaviors.

REFERENCES

Aristotle (1999). *Nicomachean Ethics* (translated by T. Irwin). Indianapolis: Hackett Publishing Company.

Dewey, J. (1975 [1909]). *Moral Principles in Education*. Carbondale: Southern Illinois University Press.

20 Resolving Ethical Lapses in the Non-Publication of Dissertations

Michael C. Roberts, Sarah E. Beals-Erickson, Spencer C. Evans, Cathleen Odar, and Kimberly S. Canter

If a doctoral dissertation is an original scholarly contribution to knowledge, then it seems important that the information gained be disseminated through means of publication, where its contribution can benefit society. In our view, for a dissertation to have an impact, submitting to ProQuest UMI, for example, does not suffice as publication, even though called "publishing" by the company. Even though this is the primary archive for dissertations and theses including an estimated "95% to 98% of all U.S. doctoral dissertations" (Proquest, 2013), limited access to these archives and their lack of use by many professionals may limit the visibility and ultimate utility of those projects only published in this forum.

Unfortunately, we have observed a number of students who have "walked away" from their dissertations (or masters' theses) after being awarded their degrees (and not just in our own graduate program). These students, for a variety of reasons, do not pursue publication of these projects. These decisions not to seek publication are rarely due to poorly conceived or inadequately conducted research. In fact, some dissertation research may be of higher quality than published research is (McLeod & Weisz, 2004). Nor have these projects resulted in uninterpretable, nonsignificant, or unoriginal findings, which therefore might not constitute uniquely valuable contributions to the scholarly literature. Indeed, in many cases, the "quality assurance" mechanisms of mentorship, committee review, and oral and written defenses are often sufficient to obviate such undesirable outcomes (McLeod & Weisz, 2004). Non-publication apparently results from a lack of further interest or incentive to publish and a desire to move on to other career activities.

To improve the rate of dissertation publishing, our graduate program transitioned to dissertations presented for final orals that are closer to the length and format required for submission for journal publication (than is the case for the more typical dissertation, with its often redundant and extraneous information). This policy improved the rate of submission of dissertations for publication because graduates did not have to extensively revise for submission, but did not fully solve the problem of graduates failing to submit and publish their projects.

We assert that dissertation non-publication represents an ethical lapse because it fails to meet the contract between the investigator and the research participant. For example, the consent form might state something to the effect that participants will give their time and effort to increase knowledge of some phenomenon and that the information will be useful for some laudatory purpose such as improvements in the understanding of human behaviors or in human-service delivery. However, the knowledge would have to be disseminated to have such a beneficial effect. Nothing in the typical consent form says: "Please participate to help me get my degree and that is all that will be gained through participation." Would very many participants, especially those from the community and clinical settings in which many students conduct their studies, agree to participate if they were told this? Participants in these settings are not receiving introductory psychology course credit and are often motivated altruistically in wanting to help people who are in a similar set of circumstances as their own. Not publishing study findings – assuming the findings are indeed publishable – is an unjustified squandering of participant efforts.

The non-publication situation raises an ethical concern because the nonfulfillment of promises to benefit others violates the contract with participants. Dowd (2004) invoked the beneficence principle of research ethics by stating: "To maximize societal benefit, results of the research must be available to society. Non-publication clearly precludes this mandate" (p. 1014).

Furthermore, failure to publish high-quality research distorts the research record on a topic of presumed importance (or perhaps it should not have been a dissertation). Importantly, publication bias may be introduced into archived research as a particular result of non-publication of completed research (e.g., Portalupi et al., 2013). This leads to concerns of the "file drawer problem" when reviewing and analyzing previous research (e.g., Pautasso, 2010). Some scientists have written about the unethical actions of investigators not publishing the results, particularly of clinical research trials (see e.g., Lehman & Loder, 2012; Yamey, 1999). However, the intentional withholding of information that may be negative or nonsupportive of a personal theory or product through non-publication or data suppression is not exactly what we are emphasizing here.

Some universities and graduate programs have recognized the possibility (even likelihood) of a student "leaving behind" an unpublished dissertation by requiring a signed agreement before final oral defense of the dissertation that states the conditions under which the student must pursue publication or relinquish some authorship rights and move down from the primary position. This agreement may state a time limit or due date by which an acceptable draft must be submitted and may also entail giving up rights to the data set if the agreement is abrogated.

Such a contract would appear to be at odds with the *American Psychological Association Ethical Principles of Psychologists and Code of Conduct*, which states: "8.12(c) Except under exceptional circumstances, a student is listed as principal author on any multiple-authored article that is substantially based on the student's doctoral dissertation." A priori agreements to publish or assign primary authorship rights to faculty could pose a problem for psychologists who are APA members and who have subscribed to the ethical code (although it may not necessarily be applicable for other academic disciplines or nonmembers). Additionally, we have been informed by our university's graduate office and university counsel that they do not condone this approach of changing authorship order to seek publication of a dissertation. The dissertation is considered copyrighted in the name of the graduate student author at the time the final document is accepted by the graduate school.

We resolved this practical and ethical conundrum by allowing teams of current students to assume responsibility for unpublished project manuscripts pulled from what the graduates self-named "the Slacker Stack." These current students on the advisor's research team complete the process of editing and submitting the manuscript based on the dissertation (and revising and resubmitting if necessary) in exchange for a junior authorship with agreement and cooperation of the dissertation author. The graduate retains the primary authorship and is involved in the preparation phase, even if not making the major effort required. Of course, the peer review process may not permit publication of all projects, but the attempt to publish needs to have been made to at least a few appropriate journals before retiring a paper. If a graduate were to forbid publication or be uncooperative in preparing the publication submission, the advisor would face an ethical challenge not yet encountered. The discussion about meeting ethical obligations would need to ensue about the rationale for non-publication as called for in the APA ethics code.

This process is understood upon joining the research team, and the graduates appear to welcome the relief from ethical guilt on non-publication. This mutually beneficial procedure has led to many more dissertations and other projects getting successfully submitted and published. Importantly, we have tried to be ethical in assisting our colleagues in maintaining their ethical contract. This approach is one way to fulfill the imperative of publishing quality research once completed.

REFERENCES

Dowd, M. D. (2004). Breaching the contract: The ethics of nonpublication of research studies. *Archives of Pediatrics and Adolescent Medicine*, 158, 1014–1015.

Lehman R., & Loder, E. (2012). Editorial: Missing clinical trial data. *British Medical Journal*, 344, d8158. doi: http://dx.doi.org/10.1136/bmj.d8158

McLeod, B. D., & Weisz, J. R. (2004). Using dissertations to examine potential bias in child and adolescent clinical trials. *Journal of Consulting and Clinical Psychology*, 72(2), 235–251.

Pautasso, M. (2010). Worsening file-drawer problem in the abstracts of natural, medical and social science databases. *Scientometrics*, 85, 193–202.

Portalupi, S., von Elm, E., Schmucker, C., Lang, B., Motschall, E., Schwarzer, G. et al. (2013). Protocol for a systematic review on the extent of nonpublication of research studies and associated study characteristics. *BioMed Central*, 2, 2. doi:10.1186/2046-4053-2-2

ProQuest. (2013). Dissertation Abstracts International. Retrieved from http://www.umi.com/en-US/catalogs/databases/detail/dai.shtml

Yamey, G. (1999). Scientists who do not publish trial results are "unethical." *British Medical Journal*, 319, 939. http://dx.doi.org/10.1136/bmj.319.7215.939a

21 Theft

Naomi Weisstein

In 1970, as a woman and an assistant professor at a little-known Catholic university, I encountered many dilemmas imposed on me from higher up in the academic hierarchy, even involving behavior toward me that violated professional ethics. One particular incident stands out in my mind. Here, in brief outline, is what happened.

I submitted a paper to a fairly young but already prestigious journal. The paper reported a dramatically novel phenomenon in the relatively new field of cognitive vision. It showed that there was a visual response in a blank area of a visual image, corresponding to whether or not that area appeared to be in back or in front of a striped grating. In other words, the visual system was computing relative depth in a blank area when there was no actual physical stimulus in that area; hence, it was responding to something in the mind.

In response to my submission, the editors asked me if I would mind them running the experiment I had described in the paper. They had "superior equipment," they said, and could do it better.

What did they mean? This could have been a legitimate request had they meant that they wanted me to add some conditions. Then why didn't they just ask me to add some conditions, or at least collaborate with them on such conditions? No. In fact, what they had said was, quite simply, that they wanted to steal my work and publish it as theirs, not mine.

I'm serious about calling it theft. If they ran my experiments, with their high rank and superior equipment, and their ownership of a journal, they would be almost sure to have a first publication of a startling effect in an up-and-coming field.

I was therefore faced with a dilemma. I could tell them yes, go ahead and rerun it. If their research worked out, I might get a citation. Excuse the vernacular, but big friggin deal! Publications trump footnotes. And women's contributions were always overlooked. It still would've been theft.

On the other hand, if I told them I didn't want them to rerun the experiment, I risked angering powerful scientists, with who knows what repercussions.

What did I do? I chose what I thought was a diplomatic middle way. I wrote back that I would be happy to collaborate with them if that's what they were proposing. I named periods of the year when I wouldn't be teaching and would be delighted to visit and work on the research.

Rapidly, I received a letter back from the editors. No, they certainly were NOT proposing a collaboration. They wrote that they only collaborated with distinguished visiting scientists. I had gotten their offer to rerun my experiment ALL wrong, they sniffed.

This could only have been written by very angry people, expressing themselves with shrieks in the manuscript. The typing was a mess. There were typos all over the letter. Also, the left-hand margins were huge, while the rightmost words were truncated.

I panicked. Their hostility was so evident and their unwillingness to have me collaborate so clear that I was sure that no matter what I did, they were going to run my experiment themselves. "I've got to publish this fast" I said to myself. "FAST." I hastily rewrote the pages to conform to the format of *Science Magazine*, a journal that published frontier findings without delay. I named five references who I knew were sympathetic to my work and I sent the new manuscript off. Then I called up the people I had named and told them the story.

Science Magazine published the article, to the accompaniment of a symphony of howls of controversy from my colleagues. While a few thought the work "wonderful," many simply disbelieved the results. An esteemed professor of visual science called the paper "notorious." A highly critical technical comment appeared in *Science Magazine* sometime after. And so on.

Eventually, the howls died down. As "cognitive vision" became more popular, I published a replication of the original work, adding more controls, and two well-respected vision scientists each replicated my findings, using two different ingenious designs. In *Science Magazine*, I published a reply to the criticisms of the technical comment with what the profession agreed were persuasive arguments supporting my research. Years later, one of the authors of the technical comment wrote me a sweet letter saying I was right all along. Nowadays, few people question the work. The experiment landed on its feet, so to speak.

The moral of this story is that, at least in 1970, the practice of science could occasionally be surprisingly corrupt. But I don't believe that the gatekeepers in those days – editors, grant panelists, tenure committee members, conference organizers – thought that they were being corrupt. Rather, I suspect that they mistakenly thought that they were advancing "excellence" in science. Thus, the editors in the incident I just described probably thought that they were promoting innovation by improving on the innovator's resources. Still other gatekeepers might have thought that they were maintaining "excellence" by excluding women, or by rejecting startling findings.

It's hard to recall how often I heard, in those days, that "women don't belong in science." And in regard to my metacontrast masking research prior to the work described earlier in this chapter, another editor remarked to a colleague that my masking functions of delay between target and mask, which were "u-shaped" (instead of the usual linear functions), were the result of my "leaning over the data." A spectacular insult, combining misogyny with rejection of unexpected results! But perhaps the editor also believed that he was maintaining "excellence."

I have written from memory about things that occurred in 1970. Are things different now? I believe they are. With the advent of Second Wave feminism, the Internet, open-access journals, journals with some of the old rules changed (e.g., a journal just announced that it will publish negative results) – with all this, the exercise of power by a few powerful men has been diluted. We should welcome such developments and work to open the gates even wider.

Science, at bottom, is a utopian enterprise. It seeks truth. It values new approaches. It cherishes reason and evidence. Indeed, it gives us some glimpse of how we might be and what we might do in a better world. Let's strive to realize that vision, and to make a fair, democratic, and equalitarian place for all who would enter that realm.

22 Claiming the Ownership of Someone Else's Idea

Dan Zakay

Sometimes a person may believe that he or she has come up with an original and innovative idea. In some cases, a literature search may show that someone else has already introduced the same idea or a very similar one. At this point, the temptation to claim ownership of the idea might be strong, and this can lead to the unethical decision to ignore or deny the existence of the original publication and to publish the idea without even referring to the original authors.

In 1985, I published an article in which a version of a classic cognitive task was introduced. There are very few versions of this classical task, and the new version was quite innovative because it increased the potential domains in which the task could be employed. Recently, an editor of a well-known journal approached me about a paper that was submitted for publication in that journal. In the paper, a version of the classical cognitive task was presented that was almost identical to the version I had published earlier. However, the original paper was not even listed in the references. One of the reviewers was familiar with the original task and with my version of it, and he drew the journal editor's attention to the fact that the original paper was not even mentioned. The editor then sent me the manuscript and asked me to comment on it.

My response was that the authors of the new manuscript were, most probably, unaware of the original publication, and I suggested asking them whether they had read it. The authors' response was surprising, if not strange. They admitted that they were very familiar with my paper and with my version of the task, but decided not to mention it because part of their manuscript involved criticizing the methodology of the original paper. This response clearly indicates that the authors made an intentionally unethical decision and deliberately ignored the original publication and thus pretended to have originated the idea. It should be emphasized that despite changes in the method of data collection, the main idea was plagiarized from my original paper. Even worse, the authors' response letter in which they explained their decision was three pages long, all of which related to my paper and my version of the task. This provides additional proof of the centrality of my task to their manuscript. The journal editor asked for a major revision of the paper. In this revised paper the authors acknowledged their intellectual debt, and I recommended it for publication.

This case is a clear demonstration of an attempt at an intellectual property theft.

23 Commentary to Part III

Susan T. Fiske

Issues of authorship and credit topped the list in our nonsystematic sample of invitees' choices to write about ethical issues in psychological science. More invitees chose this topic than any other – twice as often as the next most popular dilemma. In retrospect, this is not surprising. The commodities of our psychological science are ideas and evidence, whose ownership is harder to establish than, say, branded cattle.

Everyday misunderstandings can happen between any two people, and Michael Ross's research indicates that collaborators regularly overestimate their own contribution to a joint project. But misunderstandings are even more likely given power differentials or working across fields' distinct norms. And misunderstandings have more consequence earlier in one's career than later, as these essays indicate.

Judging from this nonsystematic sample, and from principles of psychology, people behave badly when they are subject to fewer consequences: when they are anonymous (reviewers), powerful (advisor or boss), high-status (senior or prestigious), or overloaded (all of us). Failure to give credit, idea theft, and even pressure to co-publish each illustrate these dynamics here. But sometimes subordinates also misbehave, such as trying to publish a paper using their advisor's name, without obtaining consent, or failing to cite precedents for their ideas. The current authors all agree that authorship without explicit knowledge, as well as gift authorship, is bad practice for anyone.

To be sure, assigning authorship credit is ambiguous. One example in this section includes students who leave the field, abandoning their data, which funders may still expect advisors to make public. Other examples include parallel invention, which may or may not indicate unconscious appropriation of ideas, having forgotten the source.

Several authors suggest being clear about the rules for authorship (see also the APA website). We learn that prospective authors should never assume a shared understanding, but instead agree ahead of time and appreciate that effort earns authorship. Some labs even post contracts, to give credit where it's due, respectively for the idea, design, operationalization, conducting, analysis, and write-up. In this high-stakes situation, our shared currency of authorship requires mutually shared norms and information.

Part IV

Confidentiality's Limits

24 Ethics in Service

Robert Prentky

The case I describe here presents two ethical dilemmas: (1) how to respond when confronted with a presumptively dangerous client (based solely on self-report) evaluated for an agency (limits on disclosure of confidential information in Section 4.05 of Privacy & Confidentiality in APA's Code of Conduct) and (2) how to present the case in present context while being faithful to certain details of the case and adhering to the Section 4.07 of the Privacy & Confidentiality Section.

I was a clinical director at an outpatient mental health agency in a moderate-sized New England city. One day the receptionist at the front desk informed me that a young man had been referred to the agency by an HMO for treatment of depression. All of the intake therapists were occupied, so I instructed the operator at the front desk to send him to my office. A 25-year-old Caucasian man came in and took a seat. He appeared, at first glance, well dressed and well nourished. He was remarkably articulate, but his affect was flat (and remained flat throughout the interview). At no point was there any evidence of thought disorder, delusions, or paranoid ideation. He reported no history of major mental illness. Although his current presenting problem was depression, it became evident over the course of the interview that his depression was long-standing and severe, with bouts of suicidal ideation (and several attempts).

He immediately launched into a history replete with severe childhood maltreatment and very early onset of highly unusual paraphilias (intense sexual arousal to highly atypical objects [e.g., hair or shoes], situations [e.g., exposing oneself], or individuals [e.g., children]). As he moved forward in time, he proceeded to disclose that he had a long-standing habit of going to nightclubs, meeting attractive young women, typically college students, taking them on a "romantic" walk to some private outdoor location, drugging them, and sexually assaulting them. He implied the use of one of the inexpensive, readily available street drugs that induces anterograde amnesia. He professed to not being certain whether he may have killed any of his victims. As he began talking, only in very general terms, about blatant criminal conduct, I immediately warned him about the limits of confidentiality should he disclose the details of a crime committed against an identified victim. He fully understood my warning and repeated it back to me. What began as a standard clinical intake quickly evolved into a risk analysis. Undeterred by my warning, he went on to detail a sexually violent past coincident with a long-standing, severe mood disorder, including, as noted, suicidal ideation and multiple suicide attempts (by overdosing). He appeared to have a deep reservoir of rage that alternately was internalized and expressed in suicidality or externalized and expressed as misogynistic anger.

During the three hours that I sat with him, I took copious notes and wrote up a lengthy, detailed risk analysis. In general terms (details will be omitted), the risk analysis focused on a long history of sexually violent fantasy, many reported instances of acting on his fantasies, clear evidence of planning that included manipulation, subterfuge, and drugging, and considerable, poorly managed anger that appeared primarily misogynistic. My risk report included a very clear admonition that he should *not* be treated as an outpatient. My concern was that the normal process of introspection and historical excavation could unleash emotions powerful enough to precipitate suicide or conceivably homicide. I faxed my risk analysis to the HMO, and the client was granted an inpatient evaluation at a local hospital. I was subsequently informed that he was discharged by a resident two days later. About a month afterward, I received a faxed release from a therapist in the community, requesting my risk report. I faxed the report. Several months later, a local newspaper reported that the body of a 20-year-old female had been found by a groundskeeper in a park. I immediately reported this to the CEO of the agency, who brought it to the attention of the board. I was forbidden to report my concerns to the police. I did as instructed, and there was no further resolution.

This case brought into sharp relief my responsibility to my client weighed against my responsibility to society. Since there was no Tarasoff[1] issue here – no details given about particular individuals that he reportedly victimized or intended to victimize – and no more than circumstantial evidence linking him to this victim in the park, reporting him to the police violated my ethical responsibility to protect confidential information and may have left the agency (and myself) open to a lawsuit. What limited details the newspaper provided, however, strongly suggested the client as a reasonable suspect, and my natural instinct to do what felt "right" – my moral compass as it were – exhorted me to protect society. Although it is difficult to conjecture, I suspect that if I were in private practice – not working for an agency – I would have consulted with the APA's ethics office and my liability insurance carrier about reporting the client to the local police. The downside, of course, is my recognition that it is not my job to be a police officer; it is my job to protect the confidence of my client. Thou Shall Not Breach Confidentiality is generally regarded as sacrosanct. But at what cost to society? The line that we confront may occasionally be abstruse.

When it came to writing up this case, I was confronted with yet another ethical dilemma – how to disguise sufficient details of the case while still capturing for the reader the estimated magnitude of risk posed by the client (i.e., in this case, there clearly was a signature quality to the risk that, if described, would identify the client). I elected to err on the side of caution (and ethics) and not report the substance of my risk analysis.

NOTE

1 The Supreme Court of California held in *Tarasoff v. Regents of the University of California* (1976) that mental health professionals have a duty to protect an intended victim from their patient who is threatening the victim with bodily harm.

25 Protecting Confidentiality in a Study of Adolescents' Digital Communication

Marion K. Underwood

Studying digital communication provides "a window into the secret world of adolescent peer culture" (Greenfield & Yan, 2006, p. 392). According to large, national surveys, 73% of teens (ages 12 to 17) use social networking sites (Lenhart, Purcell, Smith, & Zickuhr, 2010); 51% of teens check social networking sites daily and 22% check them more than 10 times per day (Common Sense Media, 2009). Youth ages 12 to 17 report sending an average of 60 text messages per day (Lenhart, 2012). Many teenagers claim that their social lives would end or be seriously impaired if they could not have access to text messaging (54% of girls and 40% of boys; CTIA, 2008).

This case study describes an ethical dilemma that arose in the context of an ongoing longitudinal study called the BlackBerry Project, in which adolescents were given BlackBerry devices configured to save all of their incoming and outgoing text messages to a secure online archive for later coding and analysis (Underwood, Rosen, More, Ehrenreich, & Gentsch, 2012). Adolescents and their families had been participating in a study of origins and outcomes of aggression since the third grade, and prior to starting ninth grade they were given the BlackBerry devices as a way of investigating their social aggression in the context of digital communication. Following the ethical principle of informed consent, participants and their parents knew text messages sent and received on the BlackBerries were being monitored and archived. Participants were promised confidentiality with regard to content of their digital communication, with a few important exceptions. In keeping with our ethical responsibility to protect our participants from doing harm to self or others, participants understood the limits of confidentiality – that we would alert parents and the appropriate authorities if we saw any text messaging that indicated child or elder abuse, suicidality, or intent to harm others. We obtained a Certificate of Confidentiality that protected us from being required to report other illegal behaviors; participants and parents were informed that we would not report antisocial activities, to them or to any other authorities.

The ethical dilemma arose in the summer after participants had finished their ninth grade year and had been using the BlackBerries for almost a year. A Spanish-speaking father spoke to me, distraught, trying to explain in Spanish and a few English words that his daughter had run away from home, that he had alerted the police, and that the police would be contacting me to request that I share the contents of her text messaging in the interest of locating her quickly. Because I needed some time to consider the complexities of the situation and because I was having difficulty communicating with the father, I tried to explain that I needed to see what was possible with the technology for obtaining her text messaging, because text messages were added to the online archive once every 24 hours and thus sometimes there was a delay. I also said that I would ask one of our Spanish-speaking research assistants to contact him to learn more.

Although I felt compelled to act quickly in this situation, I needed to gather some additional information and to give myself some time to think. First, I explained the situation and our responsibility to protect the daughter's confidentiality to a Spanish-speaking research assistant, who called the father within a short time and found out more about what had happened and explained to him that we wanted to help but had promised his daughter confidentiality. Second, I contacted the primary staff member and the chair of our institutional review board (IRB) for guidance, and they said that if they were contacted by the police, the university would likely urge me to release the contents of the text messaging, despite our ethical responsibilities to protect confidentiality and also our Certificate of Confidentiality.

As I considered what to do, I could not help thinking like a parent. Despite our difficulty communicating, I could hear the anguish in this father's voice. Although I knew I preferred not to share the contents of his daughter's text messaging, I immediately started thinking about whether I could do anything to encourage his daughter, our research participant, to contact him, both to ease his mind and in the interest of helping her return home. I realized that because our participant had one of our BlackBerry devices with her, I could communicate with her via text messaging, to make sure she was all right and to encourage her to contact her father. I decided that I would send her a short text message, saying who I was, and that her father was worried about her. Then I would ask her to get in touch with her father to let him know she was ok. If that was not effective, then I decided to say that I really wanted her to be able to stay in our BlackBerry Project and to keep her phone, but that if her dad got upset at me for not sharing her texts, he might withdraw his permission for her to be in the study, and then I would have to turn her BlackBerry off and she would not be able to use it anymore. I consulted with our IRB again and suggested this approach, which they supported.

In response to my initial text message introducing myself and asking if she was OK, the participant wrote back within an hour or so and said, "I am fine. I left for a reason." I then asked her if she was in any danger, wherever she was, and if she felt like she was in danger of being mistreated if she went back home, and said we could help keep her safe. Again, within an hour or so, she responded that she was safe where she was and did not wish to go home. I responded that her father was extremely worried, and asked if she could give him a call to let him know she was OK. She refused but said it was all right for me to contact him and let him know that she was safe. I asked our Spanish-speaking research assistant to share this information with the father, who was greatly relieved upon hearing the news. I then sent a text message to the participant and said that I respected that she left for a reason and that I was glad she was OK, but that it was really important to me that she be able to have her BlackBerry while she was away. I explained that I had to have her dad's permission for her to stay in the study, and if he got upset with me for not sharing her texts and telling him where she was, he might withdraw her permission to be in the BlackBerry Project, and then I would have to turn her BlackBerry service off. She quickly texted back that she would call him, which she did, and let him know she was at a neighbor's house. They were able to resolve their conflict enough for her to return home that evening.

This challenge was successfully resolved in that the participant's confidentiality was protected, I was not forced to share the contents of her text messages with the police or with her father, and she returned safely home. In responding to this ethical challenge, it was extremely helpful that she had a smartphone provided by our research project that served both as a means of my communicating with her and as an incentive for her to contact her father so she could remain in the study. I had and still have some discomfort with using the threat of her father's withdrawing her from the study and my turning off her BlackBerry service to motivate her to contact her father. It felt mildly coercive but seemed justified because it served the greater good in that it protected her confidentiality. Communicating with her in this way also seemed reasonable because what I told her was true. Because the ethical guidelines for psychologists require parental informed consent for minors, had this father become angry with me because of my refusal to cooperate in locating her, he indeed could have withdrawn his daughter from the study and I would have had no choice but to discontinue her phone service.

All things considered, I think we responded to this ethical challenge in a manner that showed respect and care for our 15-year-old participant. We were also able to respond to her father's understandable panic, while adhering to the ethical principles of confidentiality and informed consent.

REFERENCES

Common Sense Media (2009). Is social networking changing childhood: A national poll. Retrieved from the Common Sense Media Website: http://www.commonsensemedia.org/sites/default/files/Social%20Networking%20Poll%20Summary%20Results.pdf

CTIA (2008). Teenagers: A generation unplugged – A national survey by CTIA – The Wireless Association and Harris Interactive. Retrieved May 10, 2009, from http://www.ctia.org/advocacy/research/index.cfm/AID/11483.

Greenfield, P., & Yan, Z. (2006). Children, adolescents, and the internet: A new field of inquiry in developmental psychology. *Developmental Psychology*, 2006, 391–394.

Lenhart, A. (2012). Teens, smartphones, & texting. Retrieved from http://pewinternet.org/~/media//Files/Reports/2012/PIP_Teens_Smartphones_and_Texting.pdf

Lenhart, A., Purcell, K., Smith, A., & Zickuhr, K. (2010). Social media and mobile Internet use among teens and young adults. Retrieved from Pew Internet and American Life website: http://pewinternetorg/Reports/2010/Social-Media-and-Young-Adults.aspx

Underwood, M. K., Rosen, L. H., More, D., Ehrenreich, S., & Gentsch, J. K. (2012). The BlackBerry Project: Capturing the content of adolescents' electronic communication. *Developmental Psychology*, 48, 295–302.

26 Commentary to Part IV

Susan T. Fiske

All kinds of psychologists confront the limits of confidentiality – most starkly when clinical research reveals a participant's possible intent to harm self or others, as described in the contributions comprising this part. Even less clinically oriented research can present ethical challenges regarding confidentiality.

For example, a researcher using a common depression or self-esteem inventory may discover that one (anonymous) participant scores three standard deviations lower than everyone else. In our case, we decided to e-mail the entire sample, referring anyone distressed to the local mental health resources.

As another example, researchers using neuroimaging may detect incidental findings that look abnormal to a nonspecialist. In this case, the consent form must clarify that any such results would not be medically diagnostic, but that the researchers would communicate and refer the participant accordingly, unless the data were completely anonymous.

Finally, as a teacher, one sometimes encounters a disturbing essay from a fully identifiable student. Although confidentiality is not the main issue here, a personal approach, along with a referral to mental health counseling, seems in order.

Part V

Data Analysis, Reporting, and Sharing

27 Clawing Back a Promising Paper

Teresa M. Amabile, Regina Conti, and Heather Coon

TMA: Over twenty years ago, when Regina was my graduate student in psychology at Brandeis University and Heather was my lab manager, Regina and I collaborated on a creativity experiment. Regina had taken the lead on the project and was first author on the paper, which we had submitted to a journal. We were delighted when the editor wrote back after skimming the paper, saying that he found it very interesting and was sending it out for review immediately. We crossed our fingers and hoped for the best as we plunged back into our other in-progress research. This was Regina's second submission to a scholarly journal, and we were both hopeful that it would help build her curriculum vitae in the years before her job market entry. Regina excitedly shared the results at our weekly brown bag series, where Heather expressed an interest in the paper.

HC: After reading Regina and Teresa's paper, I suspected that the central analysis that they used, a True Score ANCOVA, could be useful for a study I had conducted, and so I asked Regina to show me how to do it. She agreed, and we sat down at a lab computer to look together at the syntax file for the analysis. During Regina's explanation, I spotted misplaced parentheses in the coding of the covariate. After she acknowledged the mistake, I said that I realized this could affect the paper she had under review. I felt very badly about it because I knew that Regina had invested a great deal of time in the study, and I also knew that rerunning the analysis might change the results. We talked about what to do. I did not want to pressure Regina into rechecking the analysis, but I think we both knew that this was what should happen.

RC: After Heather left, I reran the analysis with the correct coding – and discovered that the main result had vanished. It's not that the result was reversed (which could have been interesting). It was simply gone. Although the possibility of foregoing a publication was painful, I knew, from my conversation with Heather, that I had to tell Teresa about the mistake right away. I have often wondered if I would have acted as quickly if I had discovered the mistake alone. I thought about delving back into all of the analyses we had completed in an attempt to rework the paper somehow. The awkwardness of knowing that Heather might ask if I had rechecked the analysis led me to address the matter the next morning during the regular meeting that Heather and I had with Teresa. While I knew we could not let the incorrect result into the literature, I was hoping that Teresa had an idea for salvaging the paper.

TMA: That next morning, Regina walked into my office, visibly upset. Heather followed a few minutes later. Regina told me that she had something important to discuss before we talked about our current project, and then described what had happened in her reanalysis of the creativity data. I was stunned and distressed. But, after a brief discussion that made us conclude there was no way to rework the paper into something publishable, we knew that we had to call the editor immediately to withdraw it. He was disappointed, but commended us on having made the ethical decision. We knew it was, indeed, the ethical decision. Emotionally, however, it was tough. It felt like we were subjecting ourselves to a kind of claw-back – a term used often in the financial media, referring to the rescinding of compensation or bonuses that have already been paid out to executives or highly paid employees. We had been feeling that we had a nice publication practically in hand, and now it was gone.

All: In the years since this incident happened, we have told this story to our students, to illustrate the principle of integrity in reporting research results: If you discover an error in coding or analyzing data, you must reveal it to all relevant parties immediately, no matter how unpleasant the consequences – whether those relevant parties are collaborators on in-progress research, or collaborators, editors, and readers on under-review or published research. To fail to do so would be unethical, because the incorrect results would mislead anyone who would read those findings to guide their conceptualizations, future research, or practical applications; the enterprise of scientific psychology would be corrupted.

As social psychologists who understand the power of situations to influence behavior, we know that the situation worked in our favor in this episode. Would we have corrected such a mistake if one of us had discovered the error in private, if the paper had already been published, or if the error had been discovered while one of us was on the job market? We like to believe that we would have. But we recognize that our discovery and reporting of the error were due at least as much to the open and supportive collaboration we shared as it was to our individual moral fiber. And so, we teach our students to show one another their data and analyses, to think through problems together, to help one another make ethical decisions. We acknowledge that mistakes happen in research; the important thing is how we deal with them.

28 When the Data and Theory Don't Match

Bertram Gawronski

A few years ago, a study in my lab produced a pattern of results that was not only unexpected but inconsistent with a theory that my collaborators and I had proposed several years before. Making the situation even worse, the finding was directly implied by a competing theory that we aimed to refute. Our theory predicted that repeated exposure to two co-occurring stimuli would form a mental association between the two stimuli even when people reject the co-occurrence as meaningless or invalid. A useful example to illustrate this hypothesis is the concern that repeated claims of Barack Obama being Muslim may create a mental association between *Obama* and *Muslim* even when people know that the claim is factually wrong. This possibility is explicitly denied by theories assuming that newly formed memory representations depend on how people construe co-occurrences and whether they consider them as valid or invalid. Consistent with the latter theories, and in contrast to the predictions of our own theory, our study showed that the effects of repeated exposure to information about other individuals were generally qualified by the perceived validity of this information; there was no evidence for unqualified message effects that were independent of perceived validity. My graduate student and I replicated this pattern in three independent studies, so there was no question about its reliability. Yet, a major question was: What should we do with the data? Should we publish them and discredit our own theory? Or should we ignore the data and pretend that our theory is correct despite our discovery that one of its central predictions has failed?

We eventually decided to submit the data for publication, and after an initial rejection the paper was accepted pending minor revisions at another journal. It was not easy to state in the paper that our theory includes an incorrect assumption, but the data ultimately helped us better understand the phenomena our theory had been designed to explain. Since the paper came out, some people have asked me why we invested so much effort into conducting and publishing research that discredits our own theory. Looking back, I still think it was the right thing to do, because the data told us something important that was inconsistent with what we believed at that time. Two years later, someone else published a study on the same question using a different operationalization. Their results confirmed the original prediction of our theory, so it turned out that we were not completely mistaken with our initial assumptions. However, taken together, the two articles suggest that our theory is at least incomplete, in that it fails to specify an important moderator of the predicted effect (which still needs to be identified). And that's important to know if our goal is to advance science instead of pursuing our own personal agenda.

If I were in the same situation again, I would do the same thing.

Possible general principle: Admit when the data tell you that you are wrong and get them out even if you have to qualify your earlier claims.

29 Desperate Data Analysis by a Desperate Job Candidate

Jonathan Haidt

I study the ways that emotions and other motivations bias moral reasoning, and I inadvertently demonstrated the thesis while trying to prove it. I had just finished my first postdoc and had failed to get an academic job. I found another postdoc and was desperate to get more manuscripts under review at top journals before sending in the next year's applications. I had begun a line of experiments in which I exposed people to disgusting (or non-disgusting) images and stories and then measured their moral condemnation on subsequent stories. I was looking for carry-over effects of disgust.

I recruited participants in a public park in Philadelphia. The means were different across the two conditions, but the t-test was not significant because the variance was high – there were several outliers. I scrutinized those outliers carefully and realized that one of them was a guy who was smoking marijuana when I recruited him. Doesn't that justify excluding him? Maybe, but then what about the outlier on the other side, who was drinking beer while filling out the survey?

I wrestled with this problem for a while, searching for principles that would allow me to exclude the outliers that I wanted to exclude. I found a small set of principles that – with some stretching – allowed me to exclude three outliers that hurt my case while only losing one that helped me. I knew I was doing this post hoc, and that it was wrong to do so. But I was so confident that the effect was real, and I had defensible justifications! I made a deal with myself: I would go ahead and write up the manuscript now, without the outliers, and while it was under review I would collect more data, which would allow me to get the result cleanly, including all outliers.

Fortunately, I ended up recruiting more participants before finishing the manuscript, and the new data showed no trend whatsoever. So I dropped the whole study and felt an enormous sense of relief. I also felt a mix of horror and shame that I had so blatantly massaged my data to make it comply with my hopes.

Possible general principle: desperation is the mother of motivated reasoning, cutting corners, and loss of integrity. Be especially careful about your ethics when you most need or want a study to "work."

30 Own Your Errors

David Hambrick

Several years ago, after an especially grueling review process, I got a paper accepted for publication in one of the best journals in the field. It was a big deal for me. The paper addressed an important and controversial issue, and my colleague and I had spent hundreds of hours collecting the data and hundreds more writing and revising the paper. This paper was also going to be good for my career. I had gotten tenure a few years before but had hit a publication dry spell. I needed to make something happen if I wanted to stay on track for promotion to full professor. This paper was going to give me the momentum I needed. I was elated and proud. This was one of my biggest accomplishments as a researcher to date.

One thing I have learned in my twenty or so years of doing research and writing scientific papers is that errors are unavoidable. In his book *Brilliant Blunders*, Mario Livio (2013) writes that 20% of Einstein's papers contain errors. I believe it. So, as I always do when I have a paper accepted for publication, I immediately went about checking everything for accuracy. Even though I had been very careful in writing the paper before submitting it for publication, I checked every statement, every statistic, every quotation, and every reference, and when I was done, I did it again. Everything was right, and I breathed a sigh of relief. One of my worst nightmares – which will probably become one of yours if you're just getting started in research – is to have a paper accepted for publication only to discover that I have to withdraw it because of a catastrophic error.

Then it occurred to me that I should go back to our spreadsheets and check to make sure that all of the variables had been created correctly, and I'll be damned if there wasn't an error in the formula used to create one of the key variables. Fortunately, it was a small error. The correlation between the variable that we had used in the paper and the corrected one was almost perfect, and the results were off by a trivial degree. But it was an error nevertheless, and one that affected nearly every statistic in the paper.

I tried to forget about this error, but the knowledge of it gnawed at me. Fixing it wouldn't change any conclusions, but what if someone included our study in a meta-analysis someday and it *did* affect their conclusions? This wasn't likely, but it also wasn't impossible. I also remembered a story – which I call "Pearl" – that my mentor, Randy Engle, had told me about an error he made as a graduate student. The short version is that Randy spent several months running a memory experiment that called for subjects to remember lists of words, only to discover that he had made an error in the stimuli: each word was supposed to be presented once, but the word "pearl" got repeated. It was an error that Randy could have brushed under the rug, but he did the right thing: he told his advisor. He *owned* his error.

Randy's error was fatal to that experiment. The stimuli would have to be recreated, and the experiment rerun. Months of work were lost. My error was trivial by comparison – and easily fixable. So why didn't I just send the editor of the journal a simple e-mail explaining the situation (*Dear Editor: I made an error. I need to fix it.*)? Why all the handwringing? Here's why: I was scared that the editor would see this error as an indication of unforgiveable sloppiness and that my worst nightmare of having to withdraw an accepted paper would come to fruition. All of my hard work would be for naught, and that reviewer who was so unfair would dance a jig. I would be so demoralized, I imagined, that I would quit research and become a bartender. Then I did just send the editor a simple e-mail explaining the situation. I can't remember his exact reply, but the gist of it was "no problem – go ahead and fix the error." That was that.

Research is hard. Do the best you can to get everything right, and come up with a good system for checking your work. Don't rush things – I've noticed that I make errors like the one I described here when I'm in a hurry – and ask a colleague to proofread your work and serve as a second set of eyes. After all this, rest easy with the knowledge that when you do make an error, big or small, you'll own it.

REFERENCE

Livio, M. (2013). *Brilliant Blunders*. New York: Simon & Schuster.

31 Caution in Data Sharing

Richard L. Moreland

Years ago, when I was in graduate school, my adviser and I published a paper in which we described a structural equations analysis of some experimental data that we had collected. At that time, structural equations analyses were not yet common, and they nearly always focused on correlational data rather than experimental data. So there were few established guidelines to follow, and we had to rely on our own judgment, as well as the advice of some local experts.

The journal's editor, at that time, had an unusual policy that has since become standard at many other journals; he required authors to make their raw data available to anyone who asked for them, for a period of up to five years.

Soon after our paper appeared, we received a request for our data from someone who was an expert in the statistical analysis we had used, but had done no prior work in our topic area. He and his graduate student had criticisms of our statistical analysis and wanted to redo it in a "better" way, using the same data, to see if our conclusions were justified. After their new analyses were completed, the results did not seem to agree with those reported in our article, so they submitted a critique of our paper to the same journal where our paper had been published. We were given the chance to publish a rebuttal to that critique. The plan was for both papers (their critique and our rebuttal) to appear together in a future issue of the journal.

A long period of negotiation ensued, because the journal's editor was unwilling to publish any criticisms or rebuttals unless everyone involved could agree that the relevant papers were all correct. Agreement proved difficult to reach in our case, despite heated arguments on both sides (often carried out late at night, over the telephone). Eventually, however, we all decided that the two papers were as good as they were going to get, so we informed the editor that the papers were ready for publication.

For what it's worth, my own opinion is that the two papers had limited didactic value. That is, both our critics and we made some valid points – there was no clear "winner" and "loser" in the exchange. Moreover, the issues were complex and closely related to the particular data set we analyzed, so it's questionable whether the papers were very useful to readers trying to analyze other data sets of their own.

In any event, the whole experience was very stressful for me. As a graduate student, I was mortified that a paper of mine would be subjected to public criticism, but my adviser reassured me by arguing that when critiques and rebuttals are published, they do no harm to the people involved, and in fact increase everyone's interest in the original paper and the phenomenon on which that paper focused. In the end, he was right.

This whole episode shaped my attitudes toward data sharing. I became suspicious that the person who requested our data had no real interest in what we were studying, but instead just wanted to "score" a quick publication at little cost in time or energy. As I mentioned, he had no prior record of research on the topic area, and I discovered that he had published several other papers involving criticisms of other researchers' analyses related to a variety of phenomena. I wished that I had not given this person access to our data, but rather insisted that he collect his own data, of the same or similar sort. In fact, I developed a personal policy regarding data sharing that I have tried to follow ever since (although, to be honest, no one has actually asked to see my data, so no opportunities to implement the policy have arisen). My policy is as follows. If someone asks for access to my data, and those data are "special" in some way (e.g., difficult or impossible to collect again), then access will be permitted. But if someone asks for access to my data, and those data would be relatively easy to collect again, then I will deny access and instead encourage the person to collect similar data on his or her own. This would, I imagine, discourage people who are just trying to take advantage of me – people whose requests for access are not motivated by genuine interest in the relevant phenomenon.

My policy may be unrealistic, because researchers now face many pressures to share their data with anyone and everyone. For example, many granting agencies insist that data collected as part of projects they have funded must be shared, and many universities claim that data collected by their faculty and graduate students are university property and thus can be shared (or not) at the university's discretion. And I believe that the APA's ethical principles dictate that data must be shared whenever access to them is requested. Finally, recent scandals in my field regarding the possible fabrication of data by unethical researchers have created a norm that all data must be shared, so that any fabricated data can be detected.

Given all of these pressures, someone who tries to restrict access to his or her data may find it difficult to do so. Nevertheless, I would like to argue that my personal principle of selective sharing has some merit. Collecting data is very hard work, and it seems unfair to me for someone to take advantage of the work others have done for his or her own benefit. It's generally better, I think, for new data to be collected and analyzed, rather than to re-analyze old data. Maybe some guidelines can be developed that allow for the productive sharing of data, under specific circumstances and with proper acknowledgment of all the work that went into the collection of those data.

32 The Conflict Entailed in Using a Post Hoc Theory to Organize a Research Report

Thomas S. Wallsten

Kerr (1998) discusses in a very thorough and thoughtful manner the ethical dilemma that I faced many years earlier in seeking to bring a research report (Wallsten, 1976) to publication. Kerr called it HARKing, hypothesizing after the results are known. HARKing can take many forms, some of which are more egregious than others, but the essence of it is that a hypothesis, explanation, or theory developed to explain a data set or a pattern of results is introduced as motivating the experiment in the first place. Kerr (1998) documents the fact that in recent years at least some authors and research mentors have suggested, if not actually encouraged, putting ex post facto hypotheses at the front of the paper for a host of possible reasons, including that the paper becomes more readable.

We are (or at least, were) trained from graduate days on – actually for many of us, from undergraduate days on – not to present empirical research as support for a theory that grew from that very research. In broad outline form, the scientific method entails deducing predictions from theory and then testing them in an experiment. The theory accrues support if its predictions are sustained. Other possible outcomes are that the data provide partial support for the theory and the basis for revision or that they disconfirm it altogether, ideally simultaneously suggesting a new theory or perspective. The revisions or new theory are tested in a new experiment and the cycle continues. This is a very gross characterization of the ....-*deduction-inference-deduction-inference*-.... chain that links theory development and testing, but it captures the very important point that theories gain support on the basis of the outcomes of experiments designed to test them. It is misleading and a gross violation of scientific principles to develop an explanation for a set of data and then to present that data as supporting the explanation.[1]

My ethical dilemma stemmed from the fact that early in my career I was having an extremely difficult time publishing a paper that I had aimed at, and that ultimately was published in, the *Journal of Mathematical Psychology* (Wallsten, 1976). The problem was that there were so many pieces to the paper that it was hard to organize the information in a way that readers could follow. In sequential order – and the sequence is important in this story – I had developed an algebraic model of Bayesian opinion revision, ran experiments suggested by the model, and then, *post hoc*, developed an information-processing theory about the mathematical model and the empirical data. The model had its roots in conjoint measurement (Krantz, Luce, Suppes, & Tversky, 1971) and a prior study (Wallsten, 1972), not in cognitive theory per se. The model was falsifiable on the basis of empirical tests of its axioms, and it had free parameters that could be estimated under different experimental conditions. The model survived the tests, and the pattern of estimated parameter values suggested a compelling information-processing theory that pulled the entire paper together.

That is how I wrote the paper for journal submission. The introduction consisted of the conjoint-measurement model and a general Bayesian updating framework along with some motivation for the specific experimental manipulations. The method and results sections followed, and I introduced the information processing theory in the discussion. The comments of the reviewers and the journal editor were generally positive, but they all stressed how hard the paper was to follow. There were too many pieces to it and it didn't come together until the end. They would be happy to consider a revision if I could make the paper flow better and ease the readers' job.

A senior colleague who subsequently read the paper suggested that I lead off with the theory, as that gave the paper its structure. With the theory up front, all the rest of the development followed very naturally. I didn't want to do that – that is, to put the theory in the introduction and thereby suggest that it had motivated the study, which then supported the theory. But unless I did put the theory in the introduction, this work would never see the light of day.

I resolved the dilemma by presenting the theory prior to the experiments, along with a disclaimer. Specifically, in the introduction I contrasted empirical tests of formal models to empirical tests of more semantically framed theories. The former could be tested rigorously in given situations but often did not generalize easily beyond the specific paradigms. In contrast, tests of the latter often were less conclusive, but the constructs were more easily generalized to other situations of interest. Then my disclaimer followed:

It is not uncommon to obtain the advantages of both approaches by specifying a general process theory in such a way that it can be represented by well-defined formal models in particular situations. Thus, a particular model applies to specific kinds of data, but the general theory provides the interpretation of its parameters, predictions as to how the parameters will vary as a function of experimental conditions, and the means for generalizing conclusions across tasks. This interplay between model and theory forms the basis for this report, although it should be made clear at the outset that the models which follow were obtained initially as natural generalizations of the one discussed by Wallsten (1972), and that the process theory was developed after using the models to analyze the present data. (Wallsten, 1976, p. 145)

Thus, I had the advantage of presenting the unifying theory up front without the misleading implication that the subsequent data supported it. (Parenthetically, this is a solution that Kerr [1998] suggested if putting post hoc hypotheses in the introduction is necessary to facilitate good communication.) To ensure that the message remained in the reader's mind at the end of the paper, I wrote under Concluding Remarks: "Although the processing theory and the subsequent discussion are somewhat post hoc, the basic framework does have prior roots" (Wallsten, 1976, p. 183).

In retrospect, I should have omitted the word "somewhat." But it was true that the framework had prior roots, which I proceeded to outline, and at that the time I thought that fact justified the qualification. I did not outline the prior roots in the introduction because I thought doing so might have led to an impression that the specific theory came before the data, which of course it had not.

I submitted the revised manuscript and it was accepted, as I recall without much additional revision on my part. I readily admit that it still is not easy to make one's way through this paper. But the material is presented in a cohesive and logical form that the interested reader can follow, yet with intellectual and scientific honesty as to the chronology of the developments. I would suggest that when, for whatever reason, authors feel compelled to put post hoc hypotheses in the introduction, they follow Kerr's advice (include the phrase along the lines of "this hypothesis did not guide the design of this study, but rather was suggested by its results" [Kerr, 1998, p. 203]) and my example.

NOTE

1 As one example of how overly simplified this description is, sometimes theories consisting of a small number of assumptions are advanced to explain a wide range of prior experimental outcomes. It is clear upfront that the theory is subsequent to the data, but it provides a compelling explanation for them; and even then the theory is tested in new experiments. And, of course, not all legitimate empirical research is guided by well-developed theory.

REFERENCES

Kerr, N. L. (1998). HARKing: Hypothesizing after the results are known. *Personality and Social Psychology Review*, 2, 196–217.

Krantz, D. H., Luce, R. D., Suppes, P., & Tversky, A. (1971). *Foundations of measurement, Vol. I. Additive and polynomial representation.* New York: Academic Press.

Wallsten, T. S. (1972). Conjoint-measurement framework for the study of probabilistic information processing. *Psychological Review*, 79, 245–260.

——— (1976). Using conjoint-measurement models to investigate a theory about probabilistic information processing. *Journal of Mathematical Psychology*, 14, 144–184.

33 Commentary to Part V

Susan T. Fiske

Psychologists, like all humans, are subject to temptation. With regard to our data, we especially dislike having to admit to being wrong, because our reputations depend partly on our perspicacity, but also on our accuracy, integrity, and replicability of our data. We are especially tempted to ignore or deny errors if we are desperate and vulnerable, as the early-career examples presented in this Part suggest. But anyone can be tempted if the alternative – potential humiliation – is public enough and the sunk costs are large enough. The more invested we are, the more careful we must be.

Lessons in temptation here come from discovering errors at various stages of the research process: midway through a research project if one has ill-advisedly been monitoring the trends, upon analyzing results that apparently undermine a pet theory, after discovering that another theory better accounts for one's results, upon finding errors after the paper has been submitted for review, failing to replicate one's own findings, and sharing data with the risk that someone else will fail to replicate your findings with alternative analyses.

Contrary to human nature, we as scientists should welcome humiliation, because it shows that the science is working. The evidence accumulates no matter how the data will fall, human biases notwithstanding.

Part VI

Designing Research

34 Complete or Incomplete, That Is the Question: An Ethics Adventure in Experimental Design

Nancy K. Dess

Experimental psychologists love a complete factorial design. Generating data in all possible treatment combinations provides a wealth of information. Complete factorials also offer practical advantages. One is ease of statistical analysis. With off-the-rack statistical programs, results are a few mouse clicks away. Graphs and tables almost make themselves. In a graph, every filled symbol has an open-symbol counterpart, every cell in a table is full. Aesthetic virtues are elegance and symmetry. For these reasons I love factorials as much as the next experimental psychologist – but only as a starting place. My ethics tale has to do with responsible experimental design and its real-world consequences.

Design Matters

Most work in our lab over the last 20 years has concerned individual differences in taste and their correlates. A key tool is selective breeding. Selectively breeding rats for high- and low-saccharin solution intake (respectively, HiS and LoS rats) ensures robust taste differences and thus is a powerful means of examining taste correlates in, say, emotionality and social status. Years back, we wanted to know whether stress would differentially affect HiS and LoS rats' "jumpiness" (startle amplitude) and, if so, what neurochemical processes might be involved. We set about designing an experiment to find out, bearing in mind scientific objectives, practicality, and ethics. We began – of course – with a $2 \times 2 \times 2$ complete factorial design. Filling out the design would mean collecting data in eight conditions. Detecting effects with rats generally requires 8–12 observations per condition, and none of the manipulations could be repeated on the same rats. Therefore, a complete factorial design would involve roughly 80 rats.

We felt that such a design was a poor choice. On practical grounds, collecting data from 80 rats would be difficult, given the size of our colony. To streamline it, we worked through the "Three Rs" – *replacement* (using alternatives to live animals), *refinement* (minimizing pain, distress, invasiveness), and *reduction* (using fewer rats when fewer will do). These principles serve as a useful rubric for ethical decision making in research with humans, rats, or other animals (though curiously they are associated with nonhuman animal research; Perry & Dess, 2012). Could we fulfill our scientific objectives using computer simulations, in vitro methods, or other alternatives to live animals? That was easy: No. Could we fulfill our scientific objectives with fewer than 80 rats, with relatively few exposed to our stressor and/or drug treatment? That question occupied us for several meetings. Scrutinizing the complete design revealed that many comparisons bore on incidental issues. Moreover, strong a priori evidence-based reasoning made it difficult to justify including all possible stress-exposure and drug treatment groups.

Our discussions thinned the design from 8 to 5 groups, reducing the number of rats to about 50. Before committing to this plan, we reflected on how the practical, logical, and ethical advantages of this incomplete design stacked up against the pull of convention toward the complete factorial. We discussed statistical analysis, time-intensiveness, and the practical reality that an incomplete design was more likely than a complete one to draw a critical gaze. We decided that the svelte design's advantages outweighed convention's skid-greasing attraction. Finally, we revisited the specific comparisons that only the complete design would permit. Were they superfluous, or did they merit more rats? We had to confront odds we were playing based on the literature. We were omitting groups based on educated guesses about probable, but not certain, results that would reduce their value. We considered the prospect of results that would require follow-up studies – and more rats. We decided to proceed with the incomplete design on grounds that follow-up experiments were likely anyway and could be designed to clarify and extend our findings without repeating all of the original groups.

Results and Consequences

Our decisions paid off. The pattern of results allowed us to answer our central scientific questions without the omitted groups. We conducted two more experiments, to replicate and extend the findings. Both utilized incomplete designs. At the end of the three-experiment series, we were confident that we had documented an interesting, reliable phenomenon and some evidence bearing on its mechanisms. We were ready to contribute to the literature. We submitted a manuscript in a short-report format, which put us on an austerity word budget. We decided not to say much by way of justifying the incomplete designs, hoping that the logic and pattern of results would acquit them. They did not. The reviewers were not sanguine about the designs. One reviewer went so far as to say that s/he found it "difficult to believe" that we had not run certain groups. Framing the criticism as incredulity implied one of two things: The reviewer either was boggled by our apparent incompetence, or suspected that we had collected data in "missing" conditions and then omitted them from our report.

I let the former go. Although I found the incredulous reviewer's critique less than compelling, s/he was entitled to a dim view of us and/or the work. That her/his incredulity might comprise a circumspect ethics charge, however, was cause for concern. As corresponding author, after consulting with my coauthors, I shared that concern with the editor. The editor handled my concern civilly and in a manner consistent with the publisher's guidelines. The editor consulted with another editor and reported to me that they inferred bogglement (my term) rather than a misconduct allegation from the comment. The editor was receptive to resubmission after major revision. We declined, bringing the episode to a close.

Epilogue

The review process had a silver lining. We developed the manuscript into a full-length report and, as is usual, the revised manuscript benefited from some of the comments in the first set of reviews. In addition, submitting a full-length paper gave us space to explain many things more fully. This time we noted the efficiency and ethical advantages of the incomplete designs. Confident in the fit between our research questions, designs, and conclusions, we did not collect more data. The paper was published in another journal.

The first two authors on the paper were undergraduates at the time, weaving another thread into the silver lining. The project was a practicum in ethical decision making and its real-world consequences. Through designing and conducting the research, they had learned a great deal about ethics in an on-the-ground sense, a sense that revealed the inadequacy of regulatory compliance or convention as a guide to ethical decision making. In addition, irony – our carefully considered, ethically sound incomplete designs having elicited harsh criticism and innuendo – initiated a long, teachable moment that culminated with acceptance of the full-length manuscript. The lesson was not to cleave to convention, even though doing so is less likely to draw fire. (Have reviewers ever expressed concern that complete factorials included unnecessary groups?) Rather, the moral of the story was that doing the right thing seldom speaks for itself. Ethics conversations in the lab had to be shared with others for the work to have an impact. Publishing means engaging with a moral community as well as a scientific one – a lesson worth learning early.

REFERENCE

Perry, J., & Dess, N. K. (2012). Laboratory animal research ethics: A practical, educational approach. In Knapp, S., Handelsman, M., Gottlieb, M., & VandeCreek, L. (Eds.), *APA handbook of ethics in psychology, Vol 2: Practice, teaching, and research* (pp. 423–440). Washington, DC: American Psychological Association.

35 "Getting It Right" Can Also Be Wrong

Ronnie Janoff-Bulman

My student and I sat at my desk discussing the results – or rather the lack thereof – of our recent lab study. We had already conducted one experiment that provided support for our hypothesis, which we regarded as new and interesting. That first study, however, involved purely attitudinal measures, and the obvious next step was to assess actual behavior. We selected a behavioral task that appeared in the literature, made a few changes to our experimental protocol, and ran the study with ample participants per cell. Lo and behold, we found no support for our hypothesis. So there we were in my office, discussing our nonsignificant results and our next step. As good Popperians, we remained true to our hypothesis; we discussed the next study to be conducted, one that we hoped would make use of a "better" task to assess our participants' behavior – that is, a task that would produce our expected effect. This time we would "get it right."

We didn't seriously question our hypothesis; rather we scrutinized features of the experiment. We went back to the literature to seek an alternative behavioral measure and found one we believed would serve our *purpose* better. We are now conducting this revamped study. Of course the *purpose* of research should be ascertaining whether there is a true, reliable effect; but clearly our purpose was to find an effect, and in particular the effect we hypothesized. If we got it right in the current study and find our hypothesized effect, we will likely write up the two confirmatory studies, one assessing attitudes and the other behavior, and submit our paper for publication. The null effects of the intermediary experiment will not be discussed or mentioned, but will be ignored and forgotten. And null findings in this newest study wouldn't necessarily preclude revising our protocol once again to "get it right."

Now, this chapter is not an exercise in self-flagellation. Our decision to rerun our study no doubt reflects common practice in our field and science more generally. Just as it might be a mistake to ignore null results in our own research, it may also be a mistake to prematurely cede belief in one's own hypotheses. As good researchers, we have a responsibility to improve our studies and should certainly modify and attempt to perfect our manipulations and measurements. Good science is based on good research. Running a new, presumably improved study seems not only benign but laudable. After all, in part the practice reflects researchers' confidence and faith in the scientific method; that is, if done "right," research will produce "true" results, so we need to create the proper conditions for our experimental tests. If and when we find a significant effect, we believe it's real. Oh, and yes – this is the path to publication. So researchers set aside null findings and tweak manipulations and measures in search of the holy grail – the $p < .05$ result.

Unfortunately, however, this ordinary practice is not quite so benign when viewed from the larger perspective of the field. Running multiple versions of studies and ignoring the ones that "didn't work" can have far-reaching negative effects by contributing to the false positives that pervade our field and now pass for psychological knowledge. I plead guilty. And I admit that the ethical ambiguity of this common practice is really only a recent concern. With age I've become more cynical about the truth value of our research, but the recent "crisis of confidence" in our field – discussed and debated in our journals – has served as an important catalyst for my own evaluation of more-or-less ordinary research practices.

To date, most criticisms in the field have been aimed at behaviors that appear more obviously wrong. Collection of multiple dependent measures, covariates, and experimental conditions in a single study and reporting only those that prove significant, eliminating observations in analyses, adding participants to cells, and running multiple analyses to get a significant result are all practices that contribute to dramatic increases in false-positive rates. These practices are likely to be recognized as more ethically questionable or wrong than simply rerunning a study and ignoring earlier "bad" results. Although the goal is the same (i.e., getting the significant result), the link between the latter practice and the goal is not as proximal or direct; rather, there is the intervening belief that we are simply acting as good scientists to improve our research. In other words, the immediate and seemingly operative motive can be perceived as "better science" rather than "let's find a way to get a significant result." In fact, we are obligated to conduct the strongest, best studies we can and to learn from our mistakes. A major problem, of course, is that null findings may not be indicative of any mistakes.

I am reminded of an interesting set of studies conducted by neuroscientist John Crabbe and his colleagues. These researchers, who worked with mice, conducted three simultaneous and virtually identical studies in three different cities. They attempted to rigorously control every imaginable feature of the experiment; they used the exact same strains of mice and went to great lengths to equate the lab protocols, apparatus, and environmental variables. Yet despite the researchers' attempts at standardization, there were marked differences in the behavior of the mice across the labs (and reported by Crabbe and colleagues in a 1999 *Science* paper). In other words, there were large site effects, with a very large behavioral effect (the apparent outlier) evident in one lab and not the other two. Similarly, if we run and rerun studies in our labs – even if we control for as much as possible in our research with human participants (which would sadly reflect far less standardization than is possible with mice) – we would likely capitalize on chance, noise, or situational factors and increase the likelihood of finding a significant (but unreliable) effect.

So what to do? Obviously a major reason for the revamp-rerun-and-ignore exercise, like the more obvious problematic practices noted earlier, is the publication bias for significant results rather than null findings. Although utterly unlikely given current norms, reporting that we didn't find the effect in a previous study (and describing the measures and manipulations used) would be broadly informative for the field and would benefit individual researchers conducting related studies. Certainly publication of replications by others would serve as a corrective as well. A major, needed shift in research and publication norms is likely to be greatly facilitated by an embrace of open access publishing, where immediate feedback, open evaluations and peer reviews, and greater communication among researchers (including replications and null results) hold the promise of opening debate and discussion of findings. Such changes would help preclude false-positive effects from becoming prematurely reified as facts; but such changes, if they are to occur, will clearly take time.

In the meantime, there is an important role for self-awareness and self-consciousness regarding what we are actually doing as researchers. We now know that there are those among us who knowingly cheat and lie, falsifying or making up data. While an obvious threat to the legitimacy of our field, I believe these are relatively rare cases. The greater threat to the legitimacy of psychology lies in behaviors that are common – business as usual – and rarely challenged or questioned. There is value in examining our ordinary practices and recognizing the larger problems that may stem from our seemingly small behaviors. Rather than conduct our research in a self-complacent bubble, such self-examination can, over time, help improve practices. Simply recognizing and acknowledging the ethical ambiguities in our attempts to "get it right" in our research is a necessary first step in minimizing what's "wrong." This brief chapter largely reflects my own first step in this process.

36 Commentary to Part VI

Susan T. Fiske

Publishing is an implicit contract in empirical design. One essay illustrates the challenge in using incomplete designs that may not adhere to convention, but may be more ethical (less wasteful of subjects, either animal or human). Educating reviewers and other readers does burden the authors – and stretches word limits – but advancing our collective methodological and statistical practices is an often-unacknowledged but still important part of the contract.

Another feature of the implicit contract is full disclosure. Norms around this have varied from tell-the-best-story-without-lying manuscripts to provide-the-autobiography-of-the-idea manuscripts. Most often, the field falls between these extremes, although at present, the norms are shifting to recommend disclosure of all conditions, all participants, all measures, and all analyses, explaining how their inclusion or exclusion affects the reported results, and the rationale for the decisions taken. This potentially long-winded account can be available online as supplemental material, without jeopardizing the coherence of the scientific narrative. After all, coherence is another part of the implicit contract.

However, few of the current proposals for greater transparency recommend describing each and every failed pilot study. As noted, the reasons for failures to produce a given result are multiple, and supporting the null hypothesis is only one explanation. Deciding when one has failed to replicate is a matter of persistence and judgment.

What is an ethical scientist to do? One resolution is to treat a given result – even if it required fine-tuning to produce – as an existence proof: This result demonstrably can occur, at least under some circumstances. Over time, attempts to replicate will test generalizability. Ultimately, meta-analyses with moderating variables can provide some measure of faith in our findings. This gives us grounds for optimism, not just cynicism.

Part VII

Fabricating Data

37 Beware the Serial Collaborator

David C. Geary

Some time ago, I was contacted by a colleague in one of the University of Missouri's (MU) "sister" universities in another country about a potential collaborative project. I had met this person a few years earlier when he had spent a semester on sabbatical at MU, and so agreed to the collaboration. The project involved data collection from participants from four developing countries that are not typically included in psychological research, and thus the study had the potential to make a unique contribution. He stated he would organize the data collection, which would not have been particularly difficult, if I designed the study and provided all of the materials to him. Once the study was under way, we were in regular e-mail contact during which he updated me on data collection in the various locations.

Several boxes of raw data from about 500 participants arrived the following semester. This considerable amount of data was coded, double-checked, and entered for analyses by research assistants in my lab. I was eager to see the results, given the unique populations, and so, once it was ready, I focused on analyzing the data. In some ways, the results were different than I had expected, but in other ways they were not different than we had found elsewhere. In any case, the results were what they were, and I proceeded to write up the manuscript for journal review. After a few rounds of review, the manuscript was accepted for publication. I had been contacted by a colleague a few months earlier for preprints in this area, which she wanted to include in a meta-analysis on this topic. So, a few weeks after the manuscript was in press, I sent her a copy and sent a copy to another colleague who had done a considerable amount of work on the topic.

About a week or so later I received an e-mail from this second colleague telling me that their group had also collaborated with my coauthor and suspected that he had fabricated some of the data that he had collected for them in another developing country. My immediate response was "oh sh$t!" followed by a lab meeting in which I explained the situation to my staff and asked if they had noticed any unusual patterns in the data or with the protocols when they were being coded and entered for analyses. No one had noticed anything that would raise concerns. We then randomly looked through the protocols, and they appeared to have been filled out by different people; most of it was survey items and thus just required circling responses, but it did not appear as if the same person had filled them all out.

I thought about how to follow up on the concern, and after a few days it occurred to me that if the data had been fabricated, there would be similar sequences of responses across the protocols that would be highly unlikely (if the sequence was long enough) by chance or with valid protocols across different people. These sequences might occur if the same person was quickly completing multiple protocols. I wrote a short program to search for a few of these sequences – and found them! They were not with sequential protocols, however, but rather were distributed throughout the data set. We then pulled a few of the protocols with similar response patterns from the original files and these did appear to have been completed by the same person. Rather, it appeared that multiple people had completed the surveys but that each of them had completed multiple surveys that had non-sequential identification numbers and in some cases were supposedly from different countries. At this point I was convinced that some or all of the data were fabricated.

I immediately wrote an embarrassed letter to the editor of the journal that was to publish the study and requested that the article be withdrawn. The editor was very understanding, and fortunately the manuscript was only being copy edited at that point, so there was no need for a retraction of an in-print article. I then contacted the colleagues to whom I had sent the in-press manuscript, telling them that it had been withdrawn because of fabricated data. I also thanked the one colleague who alerted me to the potential problem.

The next issue was my coauthor. He didn't respond to an e-mail outlining my concerns. Given the extent to which he attempted to cover up the fabrication in the data he sent to me, and the concerns of my colleague who initially alerted me, I suspected that this was a pattern with this individual. I contacted the then-president of the Association for Psychological Sciences (then the American Psychological Society), and the ethics committee of the American Psychological Association to report the incident and to ask for options, specifically ways to alert other researchers to be wary of potential collaborations with this individual. I was told that because this individual was not a member of either organization and in fact was on another continent, there was little they or I could do about the situation. These were frustrating responses, needless to say.

In reflection, I don't think there is anything I would have done differently. In retrospect, I certainly could have vetted the raw data before proceeding with the analyses, but at that point I had no reason to suspect fraud; in fact, I found him to be very earnest and pleasant when we had met on the MU campus before our collaboration. Without a certain level of trust between collaborators, these types of projects would be nearly impossible to complete. I am, however, more cautious about committing to such projects and, for the most part, have found subsequent collaborations to be rewarding.

Nevertheless, it is important to be aware that the potential for the abuse of these relationships exists. I suspect that there are researchers who "make their living" by initiating collaborations; they may fabricate data, as in my case, as their "contribution" to the project; they may appropriate data or ideas that you generate; or they may just free-ride once the project is under way. It is probably impossible to eliminate all such risks, but it is possible to reduce them by doing a little research on collaborators. It is important to know not only about their skills and reputation in their field of expertise but also about them as colleagues. They may be at the top of their field, scientifically, but impossible to work with; a quick phone call to people with whom they have worked or to someone you know in their department will probably tell you what you need to know. I suspect that most of all you have to be aware of serial collaborators – people who have collaborated with a number of other researchers on one or two projects but cannot maintain a long-term collaboration. The serial nature of these collaborations may not be attributable to nefarious reasons, but they should alert you to potential issues with such a colleague.

38 My Ethical Dilemma

Scott O. Lilienfeld

Note that I have altered a few details of this story to prevent the individuals in question from being identified. Aside from these minor changes, the tale I relate here is factually accurate.

When I was an advanced graduate student, I served for one summer and the beginning of a Fall semester as a part-time research assistant on a long-running project. The project was headed up Dr. B, a fairly recently minted MD who had been hired by a major university's medical school. At the time, Dr. B was untenured, and was under intense pressure (which he related to me on several occasions) to publish articles in prestigious journals. Dr. B struck me as charismatic, hardworking, and ambitious. By his own admission, however, his methodological skills were not especially advanced, and he acknowledged feeling insecure as a new faculty member in a high-powered medical school environment.

One arm of the large project on which I assisted focused on neuropsychological functioning in a widely researched adult psychiatric disorder. In my role as a research assistant, I met with Dr. B on numerous occasions and became intimately familiar with the test protocol, which was administered to psychiatric inpatients and nonpatient controls. Because I had extensive background in neuropsychological assessment, I administered the test battery to a number of patients, wrote up test reports, and periodically discussed the data collection and analysis plan with Dr. B.

One day, discussing the test protocol with Dr. B, I mentioned that I was concerned that the data emanating from the project might be difficult to publish in a top-tier journal because the battery did not contain a measure of general intelligence (g). I pointed out that several of the neuropsychological tests in the battery were highly g-loaded, and that some authors had recently contended that the neuropsychological deficits presumably associated with the disorder in question were in fact attributable to g.

This alternative explanation had apparently never occurred to Dr. B. He asked whether I viewed the absence of a standard IQ measure from the neuropsychological battery as a fatal methodological flaw. I responded that I was not certain, but that he would at least need to note this omission as a limitation in the manuscript's Discussion section. As the meeting drew to a close, Dr. B seemed concerned. At the same time, he expressed hope that a decent journal would accept the article even without an IQ measure.

Approximately six months later, when I was no longer directly involved in the project, I received a phone call from a research coordinator on the project whom I had befriended. My friend was aware of my conversation with Dr. B, and he knew that the test battery did not contain an IQ measure. He called to tell me that he was alarmed by what he had just learned.

Dr. B had distributed copies of a manuscript, which he had recently submitted to a premier medical journal, to a few colleagues, one of whom shared it with my research coordinator friend (who in turn photocopied the manuscript and sent me a copy). The analyses reported in the manuscript had been based on the neuropsychological data I had helped collect.

Remarkably, the manuscript reported that a standard IQ measure had been administered to all participants. The IQ measure was described in detail in the Method section, where Dr. B mentioned that it had been included in the battery to rule out the rival hypothesis that any group differences in neuropsychological performance were due to global cognitive ability. The Results section reported descriptive statistics (mean and standard deviation) for this instrument in the sample. This section also featured the results of a multiple analysis of covariance (MANCOVA) using this IQ measure as a covariate, along with follow-up analyses, all with associated test statistics and significance levels. These analyses, which formed the centerpiece of the manuscript, revealed that the hypothesized neuropsychological differences between psychiatric patients and controls remained statistically significant even after controlling for IQ scores. In the Discussion section, Dr. B concluded that the psychiatric disorder in question appeared to be marked by specific neuropsychological deficits that could not be accounted for by IQ.

Flabbergasted by this discovery, my research coordinator friend obtained access to the SPSS file that Dr. B's statistical consultant had used to analyze the data, as this file was located on the shared laboratory computer. Although he had previously expressed a number of doubts regarding Dr. B's honesty, my friend did not want to believe that Dr. B had lied so brazenly in this case. He at first countenanced a potential explanation. Perhaps, he hoped, Dr. B had added an IQ measure to the protocol in the waning stages of the project, and Dr. B had somehow misremembered it as having been administered to all participants.

Upon investigating this possibility, my friend's worst fears were confirmed: The IQ measure was nowhere to be found in the data file. He also checked with the statistical consultant, who had no memory of having seen it. Nor did any variable name in the data file remotely resemble that of the IQ measure. My friend also discovered that the mean and standard deviation of the IQ test presented in the manuscript did not correspond to the descriptive statistics for any measure in the protocol. In an effort to leave no stone unturned, he searched all other SPSS data files on the laboratory computer to exclude the possibility that Dr. B had inadvertently reported IQ analyses derived from a different dataset. His search revealed that the IQ measure had never been administered to participants in any of Dr. B's recent or ongoing projects. After conducting a complete literature search, we further confirmed that this measure had never been used in any of Dr. B's published work.

I was dumbfounded and dismayed. I immediately contacted a fellow graduate student who had previously worked on Dr. B's research project and whom my research coordinator friend had also contacted. The two of us soon spoke in confidence with several experienced faculty members to solicit their advice. I also spoke at length with our university ombudsman. All of them recommended first raising the issue directly with Dr. B, which we did the following week in a private meeting. When we brought up our concerns, Dr. B angrily denied all wrongdoing and insisted that at worst he had perhaps committed an absentminded error while writing up the manuscript. He promised to look into the matter and get back to us promptly. For several weeks we awaited an answer, to no avail.

A well-known professor whom we trusted then encouraged us to raise the issue with Dr. B's department chair, Dr. C. In response, Dr. C called a meeting with the two of us (my fellow graduate student and I), Dr. B, and the professor whom we had approached, who agreed to serve as an informal faculty representative on our behalf. The meeting began with our outlining our allegations against Dr. B. We pointed out that no IQ measure had been administered in the project and that Dr. B was well aware of this fact, especially given that I had brought it up to him as a methodological concern. Nevertheless, the submitted manuscript contained extensive analyses using an IQ measure. We further explained that this measure was not in this data set or, so far as we could tell, any dataset with which Dr. B had been associated.

It was Dr. B's turn to respond. He maintained adamantly that he had done nothing ethically wrong, and that he had merely made an honest, albeit careless, mistake. He explained that he had once used this IQ measure in his unpublished research, and that when typing up the recent manuscript he had inadvertently cut and pasted a much older section describing this measure into the text. In response, I noted that the manuscript contained not merely a description of the IQ measure but also descriptive statistics for the measure that matched the exact N of the sample as well as detailed multivariate analyses that used the nonexistent IQ measure as a covariate. I expressed befuddlement at how these statistics could have found their way into the manuscript given that the IQ measure was not in any of the data files. Dr. B responded testily that he did not know offhand, and with a raised voice condemned us for raising questions concerning his research integrity. At that point I asked Dr. B to provide the text of the older unpublished manuscript from which the description the IQ test had purportedly been lifted. He assured us that he would do so in short order (he never did).

After hearing both sides of the story, Dr. C informed us that so far as she was concerned, the matter was officially closed. "You have both been very unfair to Dr. B," she concluded, saying that we had unjustly maligned his academic reputation and accused him of fabricating data on exceedingly flimsy grounds. The faculty representative rose to our defense, contending that we had acted courageously and in good faith, and that we were deeply concerned that Dr. B had committed academic misconduct. These words of support did not appear to assuage Dr. C, who curtly terminated the meeting shortly thereafter. A few weeks later, I learned that Dr. C had apparently put in a call to one of my research supervisors in the psychology department. Referring to me as a "troublemaker," she suggested that he might wish to entertain second thoughts about funding me on a grant project on which I recently begun to work. Fortunately for me, her advice went unheeded.

At this point I was at a loss as to how to proceed. I spoke with several well-regarded faculty members, looking for their advice regarding where to turn, and they were similarly uncertain. Dr. B's research was not federally funded, so notifying a government agency was apparently not an option. Moreover, Dr. C was a major power player in the administration of the medical school of her university, so appealing to higher authorities did not seem to be a promising avenue.

As a consequence, I reluctantly decided not to pursue the matter further. As of this writing, Dr. B is a tenured faculty member in the same school of medicine. I later discovered that the manuscript in question was eventually published in a second-tier medical journal (the journal to which it had initially been submitted had rejected it), but without the offending IQ analyses.

With the benefit of hindsight, which as we know is 20–20 (see Fischhoff, 1975), I am inclined to believe that I did not push the issue far enough. I felt exceedingly vulnerable as a graduate student and, making matters worse, I did not perceive any viable options. Neither did any of the faculty members with whom I consulted. I now wish that I had continued up the administrative ladder despite my concerns and, if necessary, requested a meeting with the dean, provost, and president of Dr. B.'s university.

Today, I regret not having done so, especially because Dr. B has continued to publish articles on the biological correlates and causes of adult mental disorders. Given that individuals who commit academic misconduct may often be repeat offenders (e.g., Errami & Garner, 2008), I am left to wonder about the trustworthiness of the data reported in these manuscripts. At the very least, this troubling episode brought home to me the vital importance of having clear-cut institutional procedures in place to address legitimate concerns regarding academic dishonesty. Had such procedures been in place at the time, the outcome might have been substantially different.

REFERENCES

Errami, M., & Garner, H. (2008). A tale of two citations. *Nature*, 451, 397–399.
Fischhoff, B. (1975). Hindsight ≠ foresight: The effect of outcome knowledge on judgment under uncertainty. *Journal of Experimental Psychology: Human Perception and Performance*, 1, 288–299.

39 Data Not to Trust

Danielle S. McNamara

Many of us trust our students to conduct a good portion of our research. Our students collect the data, collate the data, score the data, and often analyze the data. We remain as vigilant as possible to potential problems – errors can occur at any point along the way. We watch out for outliers, data entry errors, errors in the analyses, and so on, because mistakes happen. Honest mistakes and errors occur in every experiment; we're used to it. But it rarely occurs to us that the data themselves are not real; our trust in our students overwhelms the possibility that the data may be faked.

Yet, this is what happened to me early on in my career. My student (I'll call her Jane) was completing her Master's thesis. I'd recently been inspired to examine the effects of text cohesion with younger readers. Several studies had demonstrated the importance of increasing cohesion for low-knowledge adolescent and adult readers, pointing to the potential importance of cohesion for younger emerging readers. I set out on the quest to investigate this question with Jane. She was very excited about the project and decided to conduct the study for her thesis project. We decided to focus on children in approximately the second grade. We collected second grade books. We chose our passages. We developed the high- and low-cohesion versions. We constructed the comprehension questions and chose the individual difference assessments. Jane completed and defended her thesis proposal; we were all set to run the study.

Unfortunately, a devastating family crisis occurred for Jane. The crisis understandably took her away from work and consequently, progress on the study ground to a halt for some time.

To my surprise, Jane emerged one day with the fabulous news that the study was done. She had successfully collected the data, and here it was. She described how she had collected the data and presented me with a file containing the data, neatly entered into a spreadsheet.

It was quite the relief as my fears had been mounting that she might not successfully pull it together. I conducted analyses on the data, and the experiment was indeed successful – there was a pronounced effect of cohesion on students' comprehension. Students better understood the passages with increased cohesion than the original passages with low cohesion.

Where did it fall apart? Essentially, the data were too clean. The best tell was in the standard deviations. They were just too low. Certainly, there were other tells, but none that I could quite put a finger on. The data just didn't look right; it didn't feel like real data.

I asked Jane about the data. She described to me again how she had collected the data and insisted that the data were correct. She brought in the raw data and showed it to me – which, because they were second grade students, were simply Jane's transcription of their answers onto the answer sheets. What could I say?

I worried about this for some time – glancing now and again back at the data. And then it finally hit me. Playing through the study in my mind, I finally remembered what to ask for: I asked Jane for the parents' informed consent – every child needed permission from a parent in order to participate.

The sad outcome is that I never heard from Jane again after I asked for those consent forms. Jane never responded, and Jane never graduated (at least in that program under my supervision). I informed my own supervisor, but no action was taken or requested by the administration. This might seem shocking now, but I wasn't surprised at the time. Perhaps there was little to do given that she didn't seem to have any intention to continue in the program, or perhaps the severity of her family circumstances overwhelmed our sense of ethical duty. I don't know if it was ethically wrong not to pursue it further, or understandable, but that's what happened – she just went away.

My own lessons learned: I continue to watch standard deviations like a hawk. I don't think I've conducted any studies since that time with fewer than two experimenters (which would require a conspiracy to fake the data). And we always have consent forms.

To not confuse, I have since redone and published the cohesion study, but with third and fourth grade children. And, more along the lines of what one would expect with children's data, the effects of cohesion are far from pronounced and far from clean.

40 When a Research Assistant (Maybe) Fabricates Data

Steven L. Neuberg

I was a young, untenured assistant professor, a few years into my first job. Investigating social interaction processes, my laboratory studies were especially effort- and time-intensive. Each 90-minute experimental session required three participants who would engage in live interactions with at least one of the others. So that participants would be strangers to one another, research assistants (RAs) scheduled participants by phone. Multiple RAs were needed for each session to prepare the audio-recording equipment and to run participants through procedures and individual verbal debriefings. It would take a full semester – sometimes longer – to run a sufficient number of sessions to test our hypotheses. So when we couldn't schedule a full set of participants and had to cancel a session, we felt the loss; when a participant didn't show for his/her scheduled session, we felt the loss; when the audiotape equipment failed and we had to discard a session, we felt the loss. We worked very hard to minimize those losses.

One day my most trusted undergraduate research assistant – I'll call this person Julie, to protect his/her identity – came by my office, looking quite stressed. Julie told me she had stopped by the lab control room a day or so before and thought she saw another, newer RA – I'll call this person Mike – filling out a subject questionnaire, one of our main sources of data. Unfortunately, Julie had panicked and left without exploring further or questioning Mike, but, after further consideration, thought I should know what she thought she saw. Although unsure, Julie's sense was that Mike was making up or altering data. In fact, she was pretty confident this was the case, but not positive. What does one do with such information?

My first inclination was to confront Mike, but I hesitated as I worked through the implications. The best – that is, the most clarifying – outcome of such a confrontation would be a confession by Mike that, yes, he had made up some data. Such a confession would be very useful for some purposes: I could initiate disciplinary procedures, which would be a definite good for the university and for science. Yet even this "best" outcome would leave me with uncertainty regarding my data. Would I ever know that he had come fully clean with the confession? If I discarded only the data he confessed to altering, would I ever feel confident that the other data had been left untouched by his lying hands? I wasn't worried that he could have effectively generated data to inappropriately support our hypotheses: He was unaware of the specific hypotheses, they were complex (involving interactions across multiple variables), and our procedures were designed to mitigate against RAs knowing which participants had been assigned to which experimental conditions. Nonetheless, the mere creation of additional random noise for some of the variables might water down certain effects more than others and thus alter the conclusions we would draw. Even with a confession – the *best* scenario – I felt that I would never fully trust the data. The most conservative response would be to discard all the sessions up to that point in the semester.

And what if, upon questioning, Mike didn't confess – the most likely possibility – either because he was truly innocent (which I believe to this day is a decent possibility) or because he would be motivated to protect himself from the severe consequences sure to follow from a confession? What position would I be in then? She said, he said. Julie said, Mike said. I could choose to trust Julie (even though she was less than 100% confident), in which case I would need to accuse Mike formally and initiate disciplinary procedures. And I would need to throw out the data. I could choose, instead, to take Mike's expected protestations of innocence at face value and not pursue disciplinary charges – but could I ever trust the data? After all, I believed in Julie – she had been extremely competent, committed, and loyal – and, given this, could Mike ever prove to me that he *hadn't* altered the data? How do you prove such a thing? Here, too, I felt I would need to dump the data.

Regardless, then, of whether Mike would confess or not, it seemed to me that I'd have little choice but to throw away the data from up to that point in the semester. For an assistant professor in the midst of a run toward tenure, doing so would be hard. I took some solace in the sense that, if I was erring, I was erring on the side of science. But I felt the loss.

But what should I do about Mike? Should I confront him even if the outcome regarding the data would remain the same? Fabricating or altering data are damnable offenses; if Mike had indeed done so, he should be severely disciplined and his actions formally documented. With a confession from Mike, that would all be straightforward. Without a confession, however, I would be back at she said, he said, and what to do then? To move to formal proceedings with only the word of a reluctant student in hand – even one I trusted greatly – would be unconvincing. And formal proceedings would be very costly for all involved – including the great undeserved cost to Mike's reputation if Julie was wrong.

What did I do? I wasn't confident that Mike would confess. After all, he might have been innocent. And, even if guilty, I couldn't see what forces would compel him to give himself up. I suspect my decision not to confront Mike was made easier by the fact he was soon to be graduating and wasn't going into psychology or any other science; I wouldn't be unleashing a cheat into our midst. It was also made easier by my hesitation to put Julie in a difficult position – she had great potential as a scholar, and I had been nurturing her toward that end. And I had already decided to absorb the costs of all those lost data. The costs of confrontation at that point just seemed too high given the (lack of) likely benefits, so I punted. I moved him to another task that didn't involve contact with participants or data, and moved on.

Was I right not to confront Mike? With the luxury of 20-plus years of hindsight, I'm ambivalent, largely because I now have greater confidence (perhaps misguided) in my ability as an interrogator; I think I might have been able to manage the questioning in a way to minimize some of the potential costs and increase some of the potential benefits. Still, if there were to be a mistake, I think the one I'd prefer to live with is having a cheat go free than to falsely accuse on hearsay an innocent (wo)man. But the margin is tight.

I'm less ambivalent regarding the data. Throwing away a large amount of hard-earned data hurt, but "better safe than sorry" still seems like the scientifically ethical position to take. If there's to be an error – mistakenly discarding good data versus mistakenly keeping bad data – I'm still going to err on the side of mistakenly discarding good data. It's the error I'd rather live with. That said, there are an opposing set of considerations that at the time I didn't view as fundamentally moral but do now: Our participants – real people with real lives – gave their time and lent themselves to our science. I believe we are ethically obligated to respect those contributions by learning as much as we can from the data they give us. By throwing away (certainly) good data along with the (maybe) bad, I did those participants a disservice. If I could have known for sure which data were good and which data were faked or altered, I could have balanced both ethical concerns. But I couldn't. Even an ethically correct decision, viewed as a whole, may leave ethical wrongs in its wake.

41 The Pattern in the Data

Todd K. Shackelford

Some years ago, I was involved in a project in which we secured self-report surveys from collaborators in several countries. One of my roles in this project was to serve as the central repository for the hard copies of the surveys (this was before online data collection took off) and to enter the data into a statistical package. We had secured data from many different countries and cultures across many continents. What we sorely lacked were data from a country in Africa. And then we got them! I was so excited to have these data that I began entering them just as soon as they arrived in the mail. I had made it through perhaps two dozen of the 200 surveys when I thought I might have noticed a pattern in the numbers. Surely it couldn't be? I frantically flipped through 15 or 20 surveys and there it was, plain as day: the same 10-digit sequence. Our African collaborator – or someone working with him – had filled in 200 20-page surveys with the same sequence of 10-digits, over and over.

I alerted one of my senior collaborators. Neither of us could believe it. We wrestled with what we should do, how we should proceed. Should we report him? To whom? It seemed like a lot of extra work to report him, to provide the relevant evidence, and neither of us was comfortable with directly confronting him. We had never collaborated with him before and hadn't known about him until he contacted us with the offer to collect African data on whatever topic we might like to investigate. We later learned that he had contacted several U.S. academics proposing to collect African data in exchange for authorship on journal articles. It seems he had scanned recent issues of journals that published cross-cultural research and contacted authors of cross-cultural articles – casting a net and seeing what he could drag in. In the end, we decided not to confront him, but instead to alert him vaguely to "irregularities" in the data that had caused us some concern. We never heard back from him. We simply threw out his data.

In hindsight, I regret not reporting him or at least directly confronting him. This episode occurred a few decades ago, and I haven't heard anything from him or about him, or seen his name in print. But maybe

I missed it? I worry that he duped others as he attempted to dupe us, and that there is a cross-cultural project out there that has included data submitted by this person that are fraudulent, muddying the scientific literature. I have forgotten his name and university home. In fact, I have forgotten in which country he resided, and the hard copies and electronic partial files have long since been discarded. With the amount of time I have spent over the past few decades thinking about this incident and worrying whether he struck again, I surely would have been better off – and science would have been better off – had I fought through the awkwardness and dealt with this situation head on when it occurred. More specifically, I would have alerted him directly to the data fabrication, giving him the benefit of the doubt that he might have been duped, perhaps by his research assistants or local collaborators. Depending on his reaction and explanation, I would have reported the data fabrication to the head of his department or division. I might also have written to the editors of journals that publish cross-cultural work, alerting them to this instance of data fabrication. My actions would certainly have depended on his reactions and explanation. Unfortunately, I never gave him the chance to react or explain.

42 It Is Never as Simple as It Seems: The Wide-Ranging Impacts of Ethics Violations

Michael Strube

Very early in my career, I was teaching the research methods class required of our majors. The class required students to complete several demonstration research projects, each requiring collection of a small amount of data. One of the projects involved a demonstration of the illusion of control. It involved having participants express their confidence in the predicted outcome for a roll of a die – a chance-determined event over which they could have no actual control. On each of 10 rolls, participants rated their confidence in a predicted outcome, witnessed the roll of the die, and then recorded the result. The manipulations involved a 2 × 2 factorial. Some participants chose the number they thought would show up on each roll; other participants had the number predicted for them (actually the sequence chosen by a previous participant). Independently, some participants rolled the die themselves; other participants had the die rolled for them by the experimenter. The prediction was that participants would be more confident of the predicted result if they chose it themselves and rolled the die themselves – a replication of previous research. Each student in the class collected data for one participant in each of the four conditions.

When the data were assembled, one glaring problem emerged – the data set submitted by one student matched exactly the data set submitted by another. Given the number of trials involved, the probability of this actually occurring by chance are vanishingly small. I called the students in for a meeting, presented my concerns, and after first denying any wrongdoing, they finally confessed to having fudged the data (and then simply copying it for the second set). I referred the matter to our university academic integrity committee, and following a hearing of the case, that committee recommended that the students fail that particular project (I recommended to the committee that they fail the class, given the nature of the offense and class).

If that were all there was to this case, it would have been traumatic enough. It was my first encounter with blatant academic dishonesty in teaching; I found the whole affair to be emotionally draining and a real blow to my undoubtedly naïve academic worldview. But the case was much more complicated. One of the students involved was also working in my lab at the time as a research assistant. Separately from the meeting about the class data fabrication, I met with this student to discuss whether the fraud had extended to lab activities. The student swore that was not the case, but I just couldn't be sure. Before analyses were conducted on the projects, any data this student collected were eliminated from the data sets. It was not a trivial amount. The student was involved in several projects and had collected significant amounts of data for each. Some of the projects involved several graduate students as well, so progress on their research was hindered.

This episode really brought home how far-reaching ethics cases can be, even seemingly simple academic dishonesty cases. Much to my surprise, the ultimate decision about what to do with the case was out of my hands and resulted in a penalty that, to me at least, seemed little more than a slap on the wrist, especially given that the case involved research fraud in a research course. That one of the offenders was working in my lab made the case all the more painful; I felt profoundly betrayed by a student I had come to trust. And once I decided how to handle the connection of the student to data in the lab, I had to explain to other students and collaborators in my lab why the action was being taken and that completion of their projects would be delayed.

If I were in the same situation again, I would probably follow the same path. I followed university protocol for the academic integrity violation and I still think being very conservative with data management is the best practice. Now, however, I discuss research fraud and academic dishonesty more openly in classes to try and prevent these situations from occurring in the first place.

General principle: Ethical violations are rarely isolated. Their tentacles can reach beyond the particular situation to involve quite a number of other people and have far-ranging impacts.

43 Commentary to Part VII

Susan T. Fiske

Faking by a collaborator – whether long-distance or nearby, powerful or subordinate (assistant, student) – is a betrayal. Data collection requires trust: trust in the participant to follow directions and respond as faithfully as possible, trust in the data collector, and trust in the data manager. When someone violates this trust, we feel justifiably betrayed and offended. The betrayal is most acute when the fabricator is a collaborator, someone who seemed to be on the same team, sharing the same goals, but who in fact rejected those goals for more self-serving or personally expedient goals. In most instances here, the betrayal of the field as a whole, rather than personal animus, motivated the authors to confront the suspected perpetrator and notify relevant authorities, where possible – clearly the right thing to do as an individual investigator responsible for the integrity of data we report.

How can we defend ourselves as a field? The fraud was discovered in each instance, first, by downstream collaborators' attention to unusual data patterns, vigilance in investigating them, openness to reporting them, and willingness to withdraw suspect data. All these reactions are difficult because each flies in the face of human nature: We are set to believe what we see, eager to find answers, predisposed to trust collaborators, and committed to our time investments, not to mention our public pronouncements. As scientists, we must challenge our own human predispositions to accept appearances and to go with the flow.

Our collective credibility depends on our energetic vigilance.

Part VIII

Human Subjects

44 Ethical Considerations When Conducting Research on Children's Eyewitness Abilities

Kyndra C. Cleveland and Jodi A. Quas

During the past two decades, an impressive body of scientific research has emerged concerning children's eyewitness capabilities, documenting not only how well children remember salient prior experiences and how susceptible they are to errors but also the types of interview tactics that are most – and least – likely to lead to complete and accurate reports from children. Across this field, however, a recurring challenge has been how best to balance the need for ecological validity and tight experimental rigor with the ethical obligation to protect the child and family participants (see Fisher et al., 2013, for a discussion of ethics in research with children).

An example from our work illustrates how we have sought to find this balance. In 2007, we (Quas, Davis, Goodman, & Myers) conducted a study comparing children's true and intentionally false reports of body touch. Children took part in a laboratory play event with a male confederate. For one group of children, while playing, the confederate put a sticker on their neck, honked their nose, and picked them up around the waist to retrieve an item on a shelf. For a second group, he engaged in the same play activities but did not touch the children or pick them up. We thus created a salient event that included touching components and was amenable to scientific control, thereby allowing us to systematically test how well children could later recount what happened, which included answering questions about touching.

After a two-week delay, children returned and were interviewed by a social worker about the play event. During the interview, she repeatedly asked children whether they had been touched on their nose, neck, and stomach to assess the consistency of their responses. Of importance, immediately before the interview, we coached half of the children who had not been touched to falsely claim that they had been. Specifically, we asked the children to "trick" the interviewer by saying that they were touched on their noses, necks, and stomachs, explaining that we wanted to see how well we could trick her and that later we would tell her about our trick.

Our findings revealed quite remarkable abilities in children, particularly those coached to lie. In fact, the children who were lying about having been touched were more consistent in the interview than were the children who were telling the truth about having been touched. Children who were not touched and not coached to lie also performed quite well, however, indicating that spontaneous false reports of touching were extremely rare. These results, like many others coming out of scientific research concerning children's eyewitness abilities, have been quite influential in legal and policy arenas concerning how best to evaluate children's eyewitness reports (including at the level of the U.S. Supreme Court; *Kennedy v. Louisiana,* 2008) and in theoretical arenas concerning the development of mnemonic and deceptive abilities across childhood.

To evaluate the true impact of our work, though, we had to consider effects beyond the societal benefits. We had to consider the fact that we were not being completely honest with the child participants and furthermore were asking them to deceive someone else, and the impact of these facets of our study on the children themselves. We had to ensure that our procedures were not causing any harm to the children (immediately or over time), a core requirement in regulations regarding the treatment of human research participants (e.g., see section 45 CFR 46.116[d1] of the U.S. Department of Health and Human Services Federal Regulations; Section 8.08b of the American Psychological Association's [APA's] Ethical Guidelines, or the SRCD Governing Council's Ethical Standards in Research).

First, we had to make sure that the children and their parents understood and felt positive about their participation. The parents were fully informed of the purpose of the study at the start of each session. They were then able to observe the play event in its entirety and were shown the interview questions so that they could eliminate those they did not want asked (one parent eliminated three questions). The children were told about the play event and interview. They were told that they could stop either one at any time. Throughout the study, parents and children appeared positive and afterward expressed a willingness to take part in other studies.

Second, we devoted quite a bit of time to debriefing the children. This was especially critical for children who had been enticed to falsely claim having been touched. We did not want the children to come away from our study believing that deception was okay. Thus, after the interview, children were reminded of the importance of telling the truth. Children who were in the coaching condition were further told that the researcher had asked the children to trick the interviewer to see how well children could follow different kinds of instructions and whether an interviewer knew when children were tricking her or him. Of importance, the children were reminded that this was a special day and that, when an adult questions them in real life, to say only what is real. Finally, so that children did not believe the interviewer went home without knowing what happened, she returned, thanked the children, and told them they did a great job (for consistency, she did the same for children in the other conditions as well).

A third precaution in our study involved the debriefing of the parents, some of whom had watched as their child maintained an at times quite convincing false report of having been touched. We wanted to make sure parents understood that their child was following directions and their child's behavior was typical and age-appropriate. We did not want parents to be disappointed in their child for having lied, which could have affected the parents' later behavior or conversations with the child. Thus, we explained the developmental significance of children's behavior to parents and confirmed that they understood how well their child did and that their child's contribution was enormously helpful. Parents were quite receptive to our debriefing and thankful for the explanation.

In closing, when conducting research with vulnerable developmental populations, the benefits of the research must be considered concurrent with its potential for adverse effects. A clear plan for how to minimize those effects must be delineated and then followed. Overall, careful consideration of a study's design, purpose, and implications, as well as extensive debriefing (including of the parents whose children participate), can help tremendously in the pursuit of socially significant, theoretically meaningful, and ethically and developmentally appropriate research with children.

REFERENCES

Fisher, C. B., Brunnquell, D. J., Hughes, D. L., Liben, L. S. et al. (2013). Preserving and enhancing the responsible conduct of research involving children and youth: A response to proposed changes in federal regulations. *Society for Research in Child Development Social Policy Report*, 27, 1–23.

Kennedy v. Louisiana, 554 U.S. 407 (2008).

Quas, J. A., Davis, E. L., Goodman, G. S., & Myers, J. E. (2007). Repeated questions, deception, and children's true and false reports of body touch. *Child Maltreatment*, 12, 60–67.

45 Studying Harm-Doing without Doing Harm: The Case of the BBC Prison Study, the Stanford Prison Experiment, and the Role-Conformity Model of Tyranny

S. Alexander Haslam, Stephen D. Reicher, and Mark R. McDermott

The Stanford Prison Experiment (SPE) is widely recognized as one of the most ethically controversial psychology studies ever conducted. In 1971, 24 college students who had volunteered to take part in a "psychological study of prison life" were randomly assigned to roles as guards and prisoners within a "prison" that had been specially constructed in the basement of the Stanford University psychology department. As most psychology students would be aware, the study had to be brought to a premature close after six days due to the intense distress that the prisoners were experiencing at the hand of the guards. At the time, the ethical framework for conducting research of this form was poorly defined and relatively informal. But partly as a consequence of the horrors it led to, after the SPE, psychologists' code of research ethics was formalized and tightened, with the result that many felt it would never again be possible to conduct studies of this form.

Despite – or perhaps because of – this, since it was conducted, the SPE has exerted a vice-like grip over discussions about the issues of tyranny and evil that it investigated. This means that when reflecting on large-scale human atrocity, it is commonplace for researchers and commentators alike to reprise the argument that this reflects people's "natural" tendency to conform uncritically to the specifications of any group roles they are assigned, however noxious they might be. This in itself is of major ethical concern, potentially letting perpetrators off the hook. Thus, if conducting studies like the SPE raises serious ethical issues, *not* conducting them is equally of ethical concern.

The BBC Prison Experiment and a Social Identity Model of Tyranny

Thirty years later, the three of us became involved in the BBC Prison study. The first two authors designed and ran the study; the third author was the lead psychologist on an independent ethics panel that oversaw the study once it was in progress. The study was not a replication of the SPE, but used the setting of a simulated prison environment to provide a theoretically guided investigation of the key issues that Zimbardo's original study explored.

The study had two core aims. The first was to demonstrate that large-scale and impactful field studies that address powerful and troubling social phenomena can be run in ways that are compatible with contemporary ethical standards. For, like Zimbardo himself (see McDermott, 1993), we believed that an inability to investigate (and hence to understand) the bases of human inhumanity poses ethical problems in itself. But, second, at a theoretical level, we (specifically, the first two authors) were troubled by Zimbardo's role conformity explanation of his findings. We felt that by minimizing responsibility and accountability for destructive behavior, this has the potential (albeit inadvertently) to provide a warrant for future abuse – an ethical issue in itself.

These concerns were all the greater given that research in the social identity tradition suggests that people do not conform blindly to social roles, but rather enact them only when they have been internalized as aspects of a self-defining group identity. Yet while there is plenty of experimental and field research that supports this alternative analysis (and questions Zimbardo's), a key problem is that precious little of this has anything of the drama that makes the SPE so compelling – not least for those who are learning about these issues for the first time. Accordingly, when given the opportunity, we were keen to conduct research that could explore the predictions of social identity and role conformity models in a context that might match the SPE for both dynamism and scale.

Ethical Challenges and Their Resolution

As in the SPE, the basic design of our study would involve examining the behavior of men who had been assigned to groups as guards and prisoners within a purpose-built prison-like environment. Their behavior was to be video- and audio-recorded over the entire period, and to be complemented by daily psychometric and physiological measures. After the study was over, these data would be edited into four one-hour documentaries and broadcast on British television (and subsequently in a large number of countries around the world).

Clearly, before the study could proceed, it had to be subjected to a rigorous process of ethical review. This involved extended discussions with colleagues and formal submissions to both the University of Exeter's ethics panel and to the chair of the British Psychological Society's Ethics

Committee. These centered on a detailed specification of (1) the study's design and scientific goals, (2) protocols to manage risk, and (3) multiple ethical safeguards.

The design of the study is discussed in detail elsewhere (Reicher & Haslam, 2006). However, three key departures from the SPE were that (1) the experimenters had no formal role in the prison itself; (2) over the course of the study, a series of interventions were planned that were intended to increase a sense of shared social identity among the prisoners and hence their willingness and capacity to challenge the guards' authority; and (3) a series of ethical protocols and safeguards were put in place with a view to ensuring that the study caused no profound or permanent harm to participants, either physically or mentally. The latter are particularly pertinent to this volume, and so are worth enumerating.

In the first instance, potential participants underwent three phases of intensive clinical, medical, and background screening (including detailed medical and police checks) to ensure that they were neither psychologically vulnerable nor liable to put others at risk. The last phase involved extensive interviews with qualified clinical psychologists. Having been selected to participate in the study, participants signed a comprehensive consent form. This informed them that they were likely to be subjected to a range of stressors that might involve risk, including physical and psychological discomfort, prolonged confinement, and surveillance. In this regard, during the process of guard orientation, we gave the guards free rein to run the prison as they saw fit and to devise rules and punishments (including withdrawal of privileges, restricted diet, and solitary confinement) that would allow them to do this. However, they were told that they needed to respect prisoners' "basic human rights," and that (as in the SPE) physical violence would not be tolerated.

Throughout the study itself, a range of personnel were on hand to monitor proceedings and ensure participants' welfare. First, the clinical psychologists who had conducted initial screening monitored the study from start to finish. They had the right to consult any participant at any time and, if necessary, to demand that he be removed from the study. Second, a paramedic was on constant standby in case of illness or injury. Third, security guards were constantly at hand and had been issued with detailed protocols clarifying when and how to intervene in the event of dangerous behavior. Finally, an independent five-person ethics panel – all of whom had a professional interest in the issues the study addressed – monitored the study throughout. As well as the third author (a university lecturer, and chartered health and clinical psychologist), this committee was chaired by a British Member of Parliament, and also included a cofounder of Beth Shalom (an organization dedicated to genocide prevention), a council member of the Howard League for Penal Reform, and the chief advisor from the BBC's independent editorial policy unity. This committee's remit was again specified within a detailed set of protocols and included the right to demand changes to the study or, *in extremis*, to terminate it at any time (McDermott, Öpik, Smith, Taylor, & Wills, 2002). This was a critical feature that distinguished the study from the SPE.

With these various procedures in place, the study was given approval to proceed. When it did, it led to two particularly significant findings. First, participants (most notably the guards) did not conform automatically to their assigned group roles, but did so only to the extent that they actively identified with the group in question. This supports social identity theory, but is inconsistent with the claim that conformity to role is something that occurs naturally or thoughtlessly. Second, however, group identification did not entail passive acceptance of roles, but rather empowered participants to change both those roles and the system in which they were embedded (see Reicher & Haslam, 2006). Early in the study this allowed the prisoners to successfully challenge the authority of the guards; later it allowed a group of disaffected prisoners and guards to band together and conspire to institute a more draconian regime – resembling the one that Zimbardo initially created in the SPE. Rather, then, than arising from a passive process of conformity, tyranny succeeded only when those who supported it believed that what they were doing was right and hence were prepared to act with commitment. This suggests an altogether different model of tyranny than that which has dominated psychology for the last 50 years (see Haslam & Reicher, 2012).

Because this latter regime could not be implemented without violence of a form that ethical protocols would not permit, we brought the study to a premature close after eight days. Although life in the "prison" had presented many challenges for participants, up to this point no external intervention had been necessary (beyond two brief visits by the paramedic to attend to minor ailments). Now, however, we were troubled by the looming threats to order and to participants' well-being. We needed to terminate the study to ensure that "challenge" did not turn into harm. Importantly, we were reassured by both the clinical psychologists and the on-site independent ethical panel that this was the right thing to do.

After the study had ended, the participants were debriefed individually and collectively over the course of two days. Prior to the broadcast of the study's findings, all participants and the ethics committee were also given multiple opportunities to comment on rough edits of the television programs – something that led to major changes both to our analysis and to the broadcast programs. This had been a condition of ethical approval and it marked a path-breaking departure from standard editorial practices in the television industry. It reflected the fact that, contractually, the producer's role was to communicate the scientists' analysis (rather than the scientist commenting on the producer's concerns, as commonly seen in both documentaries and reality TV).

One month after the programs had been broadcast, the ethics committee published a comprehensive 14-page report detailing its own involvement in the study (McDermott et al., 2002). The summary comment that accompanied this described the study's ethical framework as "exemplary." At the end of 2005, four years after the study, we also wrote to all participants and formal observers of the study (i.e., the clinical psychologists and members of the ethics committee) and asked them to complete a detailed questionnaire reflecting on their experiences. We obtained responses from 86% of observers and 72% of participants. All indicated that their participation in the study had been a very rewarding and extremely positive experience. Although all participants had been satisfied at the time of the pre-broadcast review, we need to acknowledge the possibility that those who did not respond were less enthusiastic.

Conclusion

As we explained at the outset, our ambition in conducting the BBC Prison Study was to reinvigorate research into some of the most troubling questions of our time: When and why do humans act with inhumanity toward their peers? These questions had previously been addressed by the SPE, but, in the process, this study raised ethical concerns that had put a stop to such research. Could we address these concerns and thereby satisfy an ethical imperative to reinvestigate these questions?

For the independent observers who oversaw the study, the answer to this question was yes. Our work thus provides one template for conducting large-scale field studies that are psychologically impactful but not harmful. This has allowed us to have our say on the matter. But equally importantly, we hope that it provides a methodological platform that will ultimately allow others to have theirs. For it is only in this way that scientific debate about the psychology of tyranny can advance. And not just tyranny. Many of the "classic" studies of social psychology have recently come into question as a result of theoretical developments in our discipline (see Smith & Haslam, 2012). The challenge before us is therefore to devise field studies that match theoretical sophistication with empirical impact while also satisfying ethical principles and goals.

The stakes here are high. What the great field studies (like the SPE) managed to do was to dramatize the way that people can be transformed by the social world in which they find themselves. If we are blocked from examining such powerful social sources of variability, psychology will become increasingly dominated by the investigation of individual determinants of thought and action. This in turn will produce a skewed and misleading model of the human subject. And because human beings are reflective creatures whose actions are influenced by their own self-understanding, this must be a source of significant ethical concern.

REFERENCES

Haslam, S. A., & Reicher, S. D. (2012). Contesting the "nature" of conformity: What Milgram and Zimbardo's studies really show. *PLoS Biology*, 10(11), e1001426. doi:10.1371/journal.pbio.1001426

McDermott, M. R. (1993). On cruelty, ethics & experimentation: Profile of Philip G. Zimbardo. *The Psychologist*, 6, 456–459.

McDermott, M., Öpik, L., Smith, S., Taylor, S., & Wills, A. (2002). *"The Experiment": Report of the independent ethics panel*. London: University of East London.

Reicher, S. D., & Haslam, S. A. (2006). Rethinking the psychology of tyranny: The BBC Prison Experiment. *British Journal of Social Psychology*, 45, 1–40.

Smith, J. R., & Haslam, S. A. (Eds.). (2012). *Social psychology: Revisiting the classic studies*. London and Los Angeles: Sage.

46 Observational Research, Prediction, and Ethics: An Early-Career Dilemma

Stephen P. Hinshaw

The ethical issue I describe in this chapter originates in an early phase of my career. As an assistant professor interested in developmental psychopathology, I was eager to push the envelope of the field, conceptually, methodologically, and clinically (for recent perspectives on developmental psychopathology, see Hinshaw, 2013). A priority was devising an objective means of evaluating risk for future antisocial behavior, beyond the common strategies of asking adult informants (rater bias can be a hindrance) or giving cognitive tests (poorly validated for this purpose). Writing my first major federal grant proposal, I pondered what an advance it might be to use direct observations of child behavior for this purpose. Yet two major problems loomed: the scourge of false-positive predictions so often found in this arena, with their potentially stigmatizing labels (e.g., "pre-delinquent"); and the difficulty of creating the circumstances in which nascent antisocial behavior might be observed at sufficiently high base rates to be of use.

When all else fails, go to the scientific literature! I reread the classic work of Hartshorne and May (1928), who decades before had probed the moral character of schoolchildren. Some of the key investigations involved tempting youth to engage in behavior patterns like cheating. I pondered how to apply such methods to other forms of covert antisocial behavior, such as stealing or destroying property, already suspected as stronger predictors of delinquent behavior than overt behaviors such as physical aggression.

After much deliberation I created a brief, individual research probe, deployed during the final days of our six-week, NIH-funded summer programs for boys with attention and impulse control problems. A research assistant (RA) pulled boys from the classroom, one at a time, to participate in a study of "working on one's own." In short, for six minutes each boy was asked to be in a room by himself, with the opportunity to earn bonus points from the program's reward system by performing an academic worksheet tailored to his skill level. However, the final two problems were insoluble, answerable only if the participant looked at the answer key the RA had "accidentally" left in the workspace while giving instructions. In addition, in the study carrel, two dollar bills, three quarters, and tempting matchbox cars had been partially hidden. Magic markers and extra paper had been left as well, making it possible for the participant to deface property by marking up the room or ripping the paper.

My strong interest in behavioral observation initially led to the coding of the boy's behavior from hidden cameras. Many of the summer program's activities were video-recorded, but in this paradigm the child was not told of the hidden camera. Initial observational data were fascinating: upon seeing the temptations, some youth carefully looked around to appraise their state of aloneness and strategically placed the objects deep in their pockets. Others impulsively crammed the goods anywhere they could, with money or cars spilling out as the study ended – or scribbled on the room's walls. Still others gaped at the answer key as they filled in the answers to the final two questions.

The astute reader will see that this research scenario actually involved a double deception: (1) the use of the tempting objects and the intentionally left answer key as a kind of bait – in an investigation purportedly dealing with independent academic work; and (2) the surreptitious recording of the child's behavior. It soon became clear that simple counts (made by a different RA right after each session) of how much money was left in the room, whether property defacement had occurred, and/ or whether the insoluble problems were answered correctly provided highly reliable indicators of stealing, property destruction, and cheating. Eventually, the need for the logistically challenging and ethically questionable surreptitious video recording was circumvented.

What did the behavioral "count" data reveal? In one key study utilizing boys diagnosed with ADHD and a comparison group, we repeated the trial across two days, counterbalancing (for the clinical sample) a stimulant medication versus placebo. The results were stunning: medication significantly *reduced* stealing and property destruction by a factor of 2.5 to levels identical to those of the comparison boys, whereas cheating behavior was far *higher* (well above normative levels) on the medication day (Hinshaw, Heller, & McHale, 1992). Thus, some children lacking achievement motivation apparently chose to steal and deface property while failing to concentrate on the worksheet, yet others focused on the academic work at hand, ignoring the temptations. Clearly, medication enhanced not only such academic motivation but also the deployment of cheating to get an optimal test score!

Over time, data across multiple programs proved the psychometric value of this "temptation probe." First, counts of stealing and property destruction were correlated at extremely high levels with staff ratings of covert antisocial behavior. Second, boys with ADHD were far more likely to engage in such behaviors than were boys in our comparison group, exemplifying "known groups" validity. Third, cheating was associated with positive behaviors measured during the programs, but stealing and property destruction (uncorrelated with cheating) were linked to a host of negative correlates. Fourth, correlations of (a) observed stealing/property destruction with (b) observed physical aggression from the program's classroom and playground were far lower than similar correlations between adult ratings of covert vs. overt behavior. In short, rater "halos" might be reduced via direct observation of behavior (Hinshaw, Simmel, & Heller, 1995).

The theme of this brief chapter, of course, is ethics rather than validity. As data accumulated during the first year of the probe, I was thrilled that the task was providing unprecedented data yet simultaneously terrified about the implications of the temptation probe and the double deceit involved. Would the participants – many already at high risk for serious impairments in later life – learn that it was permissible to steal and deface property, under the auspices of a summer program intended to study them and help them? Was our research program striving to prevent criminality, or were we in fact encouraging it?

Right at that time, I moved to my present position at UC Berkeley and, of course, had to submit my entire summer program protocol for IRB review. Remarkably, my former institution had approved the temptation probe study with few questions asked. I carefully laid out the potential benefits of the temptation probe for enhancing clinical science, along with the structure and therapeutic aims of the overall program, but I also delineated its clear risks. Not long afterward, I received a message from the chair of the IRB. Half-convinced that I had been discovered as an ambitious yet potentially unethical young investigator, more interested in uncovering devious behavior patterns than protecting the rights of the participants, I approached the call with trepidation. I made sure to discuss our decision to debrief the participants in small groups (without knowledge of whether anyone had, in fact, engaged in covert behaviors) while providing a clear message as to the nature of the study and offering any participant the opportunity to privately discuss his behavior pattern with a counselor or senior staff.

After an hour's call, I emerged with a fresh perspective. The IRB chair (trained in law) actually lauded the attempt to make objective observations of antisocial behavior but asked me to consider several issues. First, the paradigm was dangerously close to entrapment, the direct enticement to engage in illegal behavior. Still, given the urgent need to understand childhood roots of such patterns and the overall tenor of the program, he believed that the committee would deem it legitimate. Second, it would be essential *not* to give parents any direct feedback about their child's response to this probe, even though we gave feedback and wrote reports on each participant utilizing nearly all other data from the program. Rather, the revised consent form would explicitly state the nature of this temptation probe, making clear that it was the sole domain from which the family could expect no information. Third, the chair reinforced the decisions to (1) cease the surreptitious videotaping, which was providing little value-added and which had comprised part of the double deceit; and (2) continue to debrief participants in small groups, providing a clear message that the program did not endorse covert antisocial actions and allowing youth to come forth individually and privately to discuss their behavior with a senior staff.

In sum, I continued the research, with the temptation probe providing important baseline observational data for prediction of later behavior patterns. The frank discussion with the IRB chair both reinforced the gravity of the paradigm's implications and assuaged my guilt over its ethical connotations, which previously had existed in a semi-vacuum.

Longitudinal research takes time. Amazingly enough, some years later our team discovered that, across all of the rich data collected in our summer programs, the single strongest predictor of the severity of adolescent male delinquency in our sample was the combined count of stealing plus property destruction from the six-minute temptation probe (Lee & Hinshaw, 2004). Intriguingly, in our subsequent work with girls showing serious attention and impulse control problems, base rates of stealing and property destruction from the probe were far lower than those of boys, with their long-term outcomes marked by devastatingly higher risk for self-harm than of antisocial behavior (Hinshaw et al., 2012).

In the end, like many colleagues, I have certainly been frustrated, from time to time, with IRB-related strictures that too often seem inimical to (rather than protective of) the conduct of legitimate research. Yet I will long remember my frank discussion with an IRB chair, the straight talk and the wisdom of which allowed me to gain respect for strong consideration of ethical issues related to observational research – in a new and extremely productive light.

REFERENCES

Hartshorne, H., & May, M. (1928). *Studies in the nature of character.* New York: Macmillan.
Hinshaw, S. P. (2013). Developmental psychopathology as a scientific discipline: Rationale, principles, and recent advances. In T. P. Beauchaine &

S. P. Hinshaw (Eds.), *Child and adolescent psychopathology* (2nd ed., pp. 1–18). Hoboken, NJ: Wiley.

Hinshaw, S. P., Heller, T., & McHale, J. P. (1992). Covert antisocial behavior in boys with attention deficit hyperactivity disorder: External validation and effects of methylphenidate. *Journal of Consulting and Clinical Psychology, 60,* 274–281.

Hinshaw, S. P., Owens, E. B., Zalecki, C., Huggins, S. P., Montenegro-Nevado, A., Schrodek, E., & Swanson, E. N. (2012). Prospective follow-up of girls with attention-deficit/hyperactivity disorder into young adulthood: Continuing impairment includes elevated risk for suicide attempts and self-injury. *Journal of Consulting and Clinical Psychology, 80,* 1041–1051.

Hinshaw, S. P., Simmel, C., & Heller, T. (1995). Multimethod assessment of covert antisocial behavior in children: Laboratory observations, adult ratings, and child self-report. *Psychological Assessment, 7,* 209–219.

Lee, S. S., & Hinshaw, S. P. (2004). Severity of adolescent delinquency among boys with and without attention-deficit hyperactivity disorder: Predictions from early antisocial behavior and peer status. *Journal of Clinical Child and Adolescent Psychology, 33,* 705–716.

47 Should We Tell the Parents? Balancing Science and Children's Needs in a Longitudinal Study

Kathy Hirsh-Pasek and Marsha Weinraub

The research described in this chapter was supported by the National Institute of Child Health and Human Development (NICHD) through a cooperative agreement (U10), which calls for scientific collaboration between the grantees and the NICHD staff. Hirsh-Pasek and Weinraub were both participating scientists in this project.

In the late 1980s and early 1990s, the United States was engaged in a massive natural experiment. More than 52% of women with children under age 6 had entered the workforce, and a new term, "dual career families," was born. In this context, families wrestled with the question of whether they should send their young children to childcare and just how much alternative care was good or bad for their babies.

At the time, the scientific community offered parents a mixed bag of advice. Some studies stressed the importance of maternal care and found that placing children in childcare was a risk (Belsky, 1999, 2001; White, 1985). Others reported the opposite effects, demonstrating cognitive and social boosts from playing with other children in childcare (Clarke-Stewart, Gruber & Fitzgerald, 1994; Lamb, 1998). Still others suggested that placing a child in childcare had little or no effect on child outcomes (e.g., Scarr, 1998). Moving from the science to more personal accusations, there were assertions that the positive results from childcare studies emerged only when women were the experimenters, and counterclaims also emerged that findings against using childcare were promulgated by researchers searching to maintain an Ozzy and Harriet caricature of the mom at home. No broad spectrum and definitive research on the topic had been conducted – until now.

The NICHD issued a request for proposals in 1987 and called for a collaborative study that would be large enough and diverse enough to address questions about the role of childcare in children's lives and development. Temple University was one of the 10 sites that earned a peer-reviewed seat at the table. Together the researchers who included voices from all perspectives, pro and con, designed a longitudinal study of children from 0 to 15 years. The study would examine children as they grew up in the rich context of their lives and would collect data at home, in childcare, and in the laboratory. Set in a Bronfenbrenner model of ecological development (Bronfenbrenner, 1986), the researchers would collect outcomes in cognition, social development, and health and would ask how various and combined contexts related to child outcomes across time at 1, 6, 15, 24, 36, and 54 months of age as well as at kindergarten, first, third, and fifth grades and beyond. With 1,364 families enrolled, the study became its own experiment in big science within the social sciences. And along the way, the sheer grandeur of the study would enable scientists to give more definitive answers to the questions of the day on the role of the home environment on parenting practices, the demographic patterns associated with amount, quality, and characteristic of childcare use, and on the variety and stability of childcare use and relationships between the use of childcare and parent-child interactions. The study would also be positioned to look not only at immediate effects of childcare but at enduring effects as well.

As is probably evident, a study of this stature and of this potential visibility offers a grab bag of ethical considerations that range from the way the questions are posed, to the nature of eligible participants, to operational definitions of constructs such as childcare. Indeed, each of these areas sparked considerable debate among the team of researchers. By way of example, one rather controversial decision was how to define what we meant by childcare. Because maternal care was and remains the norm and because we wanted to have the opportunity to study paternal care separately, a decision was made to treat paternal care like one form of childcare – alongside center care, family care, nanny care, or grandparent care. This means that when fathers regularly spent more than 10 hours a week with their child when the mother was not present, the child would be considered in childcare. This decision allowed us to observe children in the care of their fathers and was in keeping with the question of the times about the impact of non-maternal care on child outcomes, but struck some as an inappropriate way to characterize father care.

While there are a host of such issues that could be raised, here we focus on one of the many that emerge in any number of longitudinal designs of child development. This decision is one that surfaces on the ground as the data are pouring in and is one that strikes at the very core of tension between science and individual rights.

When parents and their children enlist in a study of this nature, they accept the fact that they will be living in somewhat of a fishbowl and that their data and that of their children will be aggregated to present a snapshot of trends in childcare use. Then you reach the *what if*....

What if the initial results come in and you, the researcher, see that 1 or 2 of the children out of the 136 at your site look woefully below the average of the others? What if the children who you are looking at are but 18 months of age or 2 years old, and you think that if you can help these children get language or social interventions early, you might change their developmental trajectories for the better?

Faced with this dilemma, we worried that if we did alter the course of development – or tried to – then we were not being true to the science. We were, in effect, meddling with the potential results. On the other hand, if we did not intervene, we were potentially damning a child to poor outcomes that could otherwise be reversed with careful intervention. Moreover, although the parents did not volunteer for the study to get feedback on their child's development, *what if* they did harbor concerns about the normality of their child's development? Considering us experts, the lack of a comment or concern on our part could be communicating to parents that all is OK, and our silence might deter parents from seeking the help they might otherwise have sought.

We came to a conclusion that is not necessarily the right one, but that is one that we thought we could live with. We decided to hold off on telling the parents about our concerns until we had at least one more data point that converged with our initial observations. We are not conducting clinically valid tests, so it was best to know that the child in question remained at least a standard deviation below the mean for more than one epoch in our routine assessments. Further, we could comfort ourselves in knowing that waiting for six months was not likely to harm the child, who was still quite young and potentially pliable. Indeed, at these young ages, there are no agreed-on interventions. These rationales held us until the next wave of data collection.

What if the child continued to perform at well below average levels? At that point we decided that the individual need trumped the scientific one and we would tell the parents. The number of children involved was low – approximately 2 or 3 children from our sample of 136, so it was unlikely to alter the course of the results from our site even if the parents took immediate action. And how should we inform the parents? After all, we are not physicians or clinicians, just scientists with a bit of extra data that might prove useful. We decided to write a letter to the parents in question explaining the tests that we had used, what we had found, and why we were concerned. In the letter, we suggested that the parents consider approaching their pediatrician to conduct a more thorough investigation. We also offered to speak with their primary doctor if they thought that this would be helpful.

The decision that we made came on the heels of a number of internal discussions and reasoned opinions. In the end, we recognized that we were struggling with very central principles dictated by the Belmont Report (1979), which set forth the framework for how we evaluate research with human participants. The Report calls for a balance between principles of respect for persons (treating people as autonomous individuals who acted voluntarily to participate in our study), beneficence (minimizing harm and maximizing benefits), and justice (treating people fairly). In this case, it did not seem to us that we would be fair or minimizing potential harm if we held back the information from those who could potentially use it.

The case, however, was not straightforward and could have been argued either way. Given the gravitas of the policy issues being studied, one could easily argue that letting the children develop as they would have – in their normal context – would have been of more benefit to the society at large. We decided in favor of the respect, beneficence, and justice for the individual child and family.

REFERENCES

The Belmont Report. (1979). Ethical Principles and Guidelines for the Protection of Human Subjects of Research. Retrieved Nov. 14, 2014, from http://www.hhs.gov/ohrp/humansubjects/guidance/belmont.html.

Belsky, J. (1999). Quantity of nonmaternal care and boys' problem behavior/adjustment at 3 and 5: Exploring the mediating role of parenting. *Psychiatry: Interpersonal and Biological Processes*, 62, 1–21.

(2001). Developmental risks (still) associated with early child care. *Journal of Child Psychology and Psychiatry*, 42, 845–859.

Bronfenbrenner, U. (1986). Ecology of the family as a context for human development: Research perspectives. *Developmental Psychology*, 22, 723–742.

Clarke-Stewart, K. A., Gruber, C. P., & Fitzgerald, L. M. (1994). *Children at home and in day care*. Hillsdale, NJ: Erlbaum.

Lamb, M. (1998). Nonparental child care: Context, quality, correlates and consequences. In W. Damon (Series Ed.) & I. E. Sigel & K. A. Renninger (Vol. Eds.), *Handbook of child psychology, Vol. 4. Child psychology in practice* (5th Edition, pp. 73–133). New York: Wiley.

Scarr, S. (1998). American child care today. *American Psychologist*, 53, 95–108.

White, B. (1985). *The first three years of life*. Revised edition. New York: Prentice Hall.

48 Ethics in Human Subject Research in Brazil: Working with Victims of Sexual Violence

Silvia H. Koller and Luisa F. Habigzang

This chapter focuses on the relationship between ethics committees and researchers, which is not always characterized by a spirit of collaboration. Research groups that consider sensitive topics, such as sexual violence against children and adolescents (which is the focus of our research group), often encounter critical issues that ethics committees lack the appropriate training to evaluate. Sexual violence against children and adolescents is a severe public health problem.

In 2006, the World Health Organization (WHO) and the International Society for Prevention of Child Abuse and Neglect (ISPCAN) published guidelines for the psychological assistance of victims: (1) the interventions should be structured; (2) the results should be replicable and liable to measurement and assessment; (3) the cost-efficacy results should improve over time; and (4) the interventions should apply to different contexts. In addition, the guidelines recommend that the psychotherapy approach exhibit evidence of effectiveness, such as with the use of empirical studies. The interventions should relate to the technical skills of the assisting professionals and consider the patients' individual and cultural characteristics and preferences (American Psychological Association Presidential Task Force on Evidence-Based Practice, 2006). Based on these guidelines and recommendations, we submitted to an ethics committee a proposal to assess the effectiveness (pre- and post-test design) of a cognitive behavioral intervention for girls who have been subjected to sexual violence. The study design included the application of a structured interview, based on the DSM-IV (American Psychiatric Association, 2000), to assess the symptoms of posttraumatic stress disorder (PTSD), which are common among victimized children. The ethics committee returned the proposal for revisions several times over six months. With each revision, the committee questioned a different item of the interview script. The committee eventually suggested that the entire interview be excluded, as there was concern that the process would "revictimize" the girls by provoking memories of the violence. That opinion dismissed the theoretical grounds of the project and revealed a lack of familiarity with the DSM-IV diagnostic criteria for PTSD. The opinion also displayed ignorance of the fact that diagnosis is crucial for planning appropriate interventions, which, in the proposed project, would be performed by the same team of investigators.

On scientific grounds, we wrote a letter to the ethics committee to explain that the children would be given psychological assistance and that the interview was needed to assess their PTSD symptoms and the effectiveness of the intended intervention. We explained how traumatic memories work, and we described the need for psychotherapeutic interventions that focus on restructuring memory fragments to alleviate symptoms. We described the positive results of our research group's previous studies with the same interview, and we grounded the project in previous literature.

The ethics committee responded by saying that the project would be approved as long as the interview was excluded. The response indicated that the investigators "ought not mention the sexual abuse in the interviews with the children." The committee does have grounds to be concerned, as leading or poorly conducted interviews can produce negative effects on research subjects. However, in the case of sexual violence, the traumatic element is the actual experience of abuse rather than talking about the abuse. Using therapeutic methods to talk about the experience of violence and applying effective intervention strategies with trained and qualified professionals can reduce children's symptoms and improve their quality of life.

Following the recommendations of National Research Ethics Committee, we sent the proposal to our university's Central Ethics Committee, which immediately approved the project. The study was performed, and the results showed that the intervention was effective at reducing symptoms of depression, anxiety, and PTSD (Habigzang et al., 2009, 2013). In addition, the study results were used to design a professional training program for psychologists who assist victims in the public health care system. With funding from national research support agencies, psychologists have been trained and have provided positive assessments of the intervention and training program (Habigzang et al., 2010). That protocol is currently the only available protocol in the country that exhibits evidence of effectiveness.

The use of previous informed consent is an additional polemical issue in research on sexual violence. It is required that the parents or legal guardians of children sign the informed consent form before they can be authorized to participate in studies. For this purpose, the study aims must be clearly described. Actually, the process of informed consent is separated into two steps. First it is explained what the research is and what it concerns, its confidentiality and its limits, and a document has to be signed (legalistic paper). So the procedures are explained to the participants before the interview is started. As a second step, after the data collection, participants sign consent forms to allow the researchers to use their data. Debriefing (explaining more, reassuring them, etc.) seems useful in cases of intrafamilial violence. Family members still may decide if they do not want the researchers to use their data, provide consent, and allow the proper authorities to be notified when children and adolescents require protection. For example, our group has noted that abusive parents or legal guardians often refuse consent when they realize that violence is the focus of the study. As a result, the families in which violence occurs are not investigated, and the children continue to be victimized until the protection agencies are alerted. Therefore, the at-risk children are not included in studies because the abusive caretakers do not provide consent.

The use of informed consent to clarify the aims of research, after the interview, may simplify the process of working with families that may not provide previous consent, assist with data collection, and allow the proper authorities to be notified when children and adolescents require protection. Without debriefing, informed consent can be a barrier to research. For example, our team encountered a case that involved a half-illiterate mother of a girl who had been subjected to sexual violence. When the mother was presented with an informed consent form that was written with simple and objective language, she denied authorization. The mother thought that the consent form was written in overly simple language; she indicated that she would have provided consent if the consent form had used sophisticated terms that she could not understand. According to her, the use of sophisticated language would ensure that the study was university-based. Because the form was written using simple language, she was wary that the study was being conducted by the police.

These cases show us that we have much to learn about the relationships among real life, research, and ethics. Appropriate training for ethics committee members and scholars should facilitate the dialogue with the individuals who subject themselves to our protocols and resolutions. The analysis of a research project should not be based on the theoretical views of an ethics committee, and approval should be given when the studies are based on scientific grounds that justify the research with the intended methods, given that the methods comply with the ethical regulations. Theoretical and political considerations and congeniality between committee members and project proponents may hinder the development of science. The result may include the delay of benefits that study results can provide to participants and other people with similar characteristics.

REFERENCES

American Psychiatric Association. (2000). *Diagnostic and statistical manual of mental disorders* (4th ed., text rev.). Washington, DC: APA.

American Psychological Association Presidential Task Force on Evidence-Based Practice. (2006). Evidence-based practice in psychology. *American Psychologist*, 61(4), 271–285.

Habigzang, L. F., Damásio, B. F., & Koller, S. H. (2010). Training program for mental health professionals with a focus on treatment of victims of sexual abuse: A social technology developed in Brazil. In ISPCAN (Ed.), *ISPCAN World Perspectives 2010* (pp. 345–348). New York: ISPCAN.

(2013). Impact evaluation of a cognitive behavioral group therapy model in Brazilian sexually abused girls. *Journal of Sexual Abuse*, 22(2), 173–190.

Habigzang, L. F., Stroeher, F., Hatzenberger, R., Cunha, R., Ramos, M., & Koller, S. H. (2009). Grupoterapia cognitivo-comportamental para crianças e adolescentes vítimas de abuso sexual [Cognitive behavioral group therapy for children and adolescents victims of sexual abuse]. *Revista de Saúde Pública [Journal of Public Health]*, 43(1), 70–78.

World Health Organization & International Society for Prevention of Child Abuse and Neglect. (2006). *Preventing child maltreatment: A guide to taking action and generating evidence.* Geneva: World Health Organization.

49 Honesty in Scientific Study

William B. Swann

A colleague asked me whether a project he was considering doing was scientifically and ethically viable. The study was designed to replicate (and extend) a published study that had been conducted using Amazon's *Mechanical Turk*. In the original study, participants received small amounts of money as an incentive for participating. My colleague did not have money to pay participants but he could offer them the chance of winning an iPod as an incentive. He recognized, however, that offering a *chance* to win an iPod was not the same as offering a *certainty* of acquiring money, and he worried that readers would insist that he had not fully replicated the earlier study. To address this concern, he suggested telling participants that they would receive money for participating "as long as they solved a problem" that they would receive after performing the other components of the experiment. The "problem" would be insolvable, so he would not be obligated to pay them money (which he didn't have), but in the spirit of compensating participants, he planned to enter them into a lottery for the iPod after they were told that they had not solved the problem.

He reasoned that the study was ethical for two reasons. First, technically he did not lie to them when he told them that he would give them the money if they completed the problem (although he would be lying by omission become most if not all participants would assume that the problem was solvable). Second, because he intended to award someone an iPod, he did intend to compensate participants.

I told him that there were scientific as well as ethical difficulties with his plan. First, by adding the contingency that participants would be paid only if they solved the problem, he was changing the experimental design enough that it could no longer be considered a replication. Second, the study was unethical because telling participants that they would receive money if they solved a problem that was in fact unsolvable was not only a lie of omission; it was an *unnecessary* lie. The general principle here is that lying to participants is permissible only when there is no alternative. In the study proposed by my colleague, lying was an expedient that served to accomplish an aim that could have been accomplished in numerous alternative ways, such as by teaming up with a collaborator who had research funds, applying for a grant, and so on. If approached with a similar question again, I would have the same response.

50 Ethically Questionable Research

William von Hippel

In early November 1999 I joined Stormfront, one of the first and largest neo-Nazi sites on the Internet. Every morning for about a year I logged into various discussion forums on the site to see what my Nazi cyber-friends were doing and how they were reacting to the events of the day. Soon I also started visiting the websites of White Aryan Resistance and the World Church of the Creator (proponents of Racial Holy War). What interested me about these groups was how the virulent racism of their members might differ psychologically from the everyday sort of prejudice that I typically studied in the laboratory.

Perhaps not surprisingly, one notable difference was the pride, conviction, and vindictiveness with which group members declared the inferiority of other races to their own, and indeed reveled in the hate crimes the reports of which occasionally appeared in the news. For example, although the murder of James Byrd took place more than a year prior to my joining, it was still a major topic of conversation among group members. One member even set up a "nigger dragging" contest on his website – complete with animations – to see who else could drag an African American to his death from the back of a truck but somehow keep the body intact for longer than had been the case with Byrd's killers. My colleagues and I wondered how people developed such extraordinary attitudes, so we decided to conduct an experiment with group members to see whether these neo-Nazis were more motivated by out-group denigration (as seemed to be the case) or by in-group favoritism (as they claim in their discussions; Stormfront's motto is "White Pride Worldwide").

To examine this question we presented members of Stormfront with out-group-derogating and in-group-enhancing messages. We assessed how much interest the different types of messages generated by measuring the number of reads and responses for each message. Because responses to the messages would likely be influenced by group members' knowledge that we were researchers, we thought it important to pretend that we were genuine group members. Additionally, because the actual messages sent by group members were virulently racist, it was important that our own messages matched the tone of those sent by others. Lastly, we could not conceive of a manner in which we could debrief participants without upsetting them.

After debating the various ethical issues among ourselves and discussing our research protocol with our IRB (which approved the research after extensive deliberations), we decided to proceed with the project. I had long since joined the group under false pretenses, and with the pseudonym I had created we sent out a series of messages over a period of several months. As we hypothesized, group members were more likely to read and respond to out-group-derogating messages than to in-group-enhancing messages. These findings provide tentative evidence that intergroup threat might be an important source of prejudice among these Internet Nazis (for further details, see Gonsalkorale & von Hippel, 2012). After completing the experiment, we stopped posting to the group and essentially disappeared from group life.

To summarize the ethical concerns, we conducted an experiment on people who did not consent to their involvement or even know that they were involved in a research project. We deceived them about our attitudes and behaviors (i.e., we posted false stories), and then we failed to debrief them when the study ended. Additionally, our messages contributed to the hate speech that filled the site, and thus might at some level have led to a hardening of group members' attitudes.

Should we have been allowed to conduct this research? I am confident that our ethics request would have been denied if we had attempted to conduct a similar project among a more sympathetic group, such as a support group for parents of sick children. I also wonder whether we should have debriefed participants. If a research project is sufficiently upsetting that debriefing would make matters worse, then the project probably shouldn't be conducted. In this case the IRB concurred with our judgment at the time that we shouldn't debrief, but I'm no longer sure that this was the right decision. I'm also not convinced we should have conducted the project in the first place, so I don't know whether I'd do the same thing if the opportunity arose.

REFERENCE

Gonsalkorale, K., & von Hippel, W. (2012). Intergroup relations in the 21st century: Ingroup positivity and outgroup negativity among members of an internet hate group. In R. Kramer, G. Leonardelli, & R. Livingston (Eds.), *Social cognition, social identity, and intergroup relations: Festschrift in honor of Marilynn Brewer*, pp. 163–188. Boston: Taylor & Francis.

51 Commentary to Part VIII

Susan T. Fiske

Ethical treatment of human research participants boasts a huge literature, most recently the National Research Council report on Revisions to the Common Rule for the Use of Human Subjects, a panel chaired by yours truly. Revisions to the Common Rule aim to preserve ethical treatment while reducing the burden on investigators and institutional review boards (IRBs). Core principles of ethical treatment include the Belmont Report's respect for persons, beneficence, and justice.

Nowhere are these principles more important for behavioral and social sciences than in cases of investigating the evils of human behaviors such as those confronted in this Part: child abuse, power exploitation, sexual violence, cheating, lying, and intergroup hatred. As a science, we cannot understand social issues and make the world a better place without tackling uncomfortable topics. So beneficence and justice (and often respecting the autonomy of persons) all require that we as a science and as a larger ethical community confront the trade-offs in this potentially controversial research. We have an ethical responsibility to promote this kind of research.

Reducing burdens on investigators and IRBs is an ethical responsibility that promotes potentially beneficial research and focuses IRB expertise on cases where it truly matters to participants' welfare. Human research ethics must balance the promotion of research and its benefits to humanity with the prevention of harm to participants; both are moral obligations.

Part IX

Personnel Decisions

52 Culture, Fellowship Opportunities, and Ethical Issues for Decision Makers

Richard W. Brislin and Valerie Rosenblatt

An ethical dilemma faced by the senior author dealt with the awarding of fellowships for advanced graduate students and recent recipients of doctoral degrees. (The second author will give reactions based on her experiences both as an international student and as a recent PhD.) I was working at the East-West Center, an international research and educational organization in Honolulu, Hawaii, which is housed on the University of Hawaii campus. In different years since its inception in 1961 it has been part of the University of Hawaii and has been a separate institution that has received most of its funding from the U.S. Government. One of its major tasks has been to provide opportunities for educational, technical, and cultural exchanges for citizens of the United States and of Asian and Pacific Island nations.

I had the opportunity to offer year-long predoctoral and postdoctoral fellowships to scholars who were interested in pursuing cross-cultural research studies. The fellowships were for citizens of any country in the East-West Center mission statement: Japan, Korea, Malaysia, India, Pakistan, Taiwan, Indonesia, Australia, New Zealand, New Guinea, and so forth. During the good years in the 1970s through the 1990s, I could have two fellowships to offer each year. But East-West Center rules were careful to avoid the perception of favorite countries, favorite institutions, or "old boy" networks. I could not simply write to a colleague in Japan and ask her or him to send me a first-rate scholar. I had to prepare a general announcement that would go to institutions in all of the countries in the East-West Center mission statement.

Thus, a variety of professors and government officials would receive the announcement. They could then choose to nominate someone or choose not to do so. The problem stems from the situation into which the high-status nominators are being placed. They have to ask scholars if they would like to be nominated. Hawaii is a great place to spend a year, and so many people said "yes." Let's use the example of professors and postdoctoral scholars in Asia. If professors nominate someone, they expect that person to be given the award. If the nominee does not receive an award, the professors may feel insulted that their status is being challenged and consequentially feel a loss of face. The professors become angry with the East-West Center and with their colleague who requested nominations. I did not enjoy putting Asian professors into these situations with the accompanying emotional experiences, and hence I felt that ethical guidelines were being challenged.

First author's reaction: The problem stems from a cultural difference. (I discussed such differences as they apply to workplace issues in Brislin, 2008.) Professors in Asia receive more deference from students and younger scholars than do professors in the West. Some American and Canadian readers of this case study may be able to relate to this difference. If they have had a sabbatical or a guest lecture tour in Asia, students probably have stood when they entered a room. Students take notes to the point of hanging on every word the visiting professor offers. Students rarely ask questions out of concern that they might be seen as challenging the status of the speaker.

So how did I deal with this problem? My approach was probably imperfect. In my nomination requests and other communications with potential nominators (made easier today with e-mail) I would appeal to treatments of cultural differences. These include the work of Hofstede (2001) and House et al. (2004). I pointed out that many Asian countries are higher on the power distance dimension than the United States is. Powerful and high-status people (and this includes professors) in high-power-distance countries receive more deference and have more unchecked decision-making power than do individuals in low-power-distance cultures. Asian professors are far less accustomed than American professors are to having their recommendations seemingly ignored.

Many Asian countries are also higher in uncertainty avoidance than the United States is. This means that there are more norms and that the norms are taken more seriously (Gelfand, 2012). Norms guide everyday behaviors. In the case under discussion, the normative method of selecting fellows is that a high-status professor nominates someone. That "someone" is then chosen. I might add that part of this normative practice is that younger scholars court the favor of senior professors over a period of many years so that they are the "someone" who is nominated for awards.

Issues were made more complex by another aspect of decision making surrounding the awarding of fellowships. Countries high in power distance and uncertainty avoidance can have complex bureaucracies. Officials take their power seriously, and norms evolve that protect the turf of decision makers. One way to cut through red tape is to call on well-established interpersonal relationships. Asian officials can say to themselves, "This guy has told me about the decision making process, but he'll downplay it for me." This allows me to cite Monty Python. The officials who nominate potential fellows assume that I am engaging in "wink-wink, nudge-nudge" when I describe the decision-making process.

Problems such as those described here will not go away easily. People with resources to offer (travel grants, fellowships, award extensions) can address the issues with one of the basic guidelines for ethical professional practice. They can try to encourage informed consent surrounding all the issues involved in the decision to participate in an endeavor, or they can choose nonparticipation.

Second author's reaction: As an immigrant and a former international student who had an opportunity to study and work in Australia, Singapore, Russia, and the United States, I also see cultural differences, especially those pertaining to personal networking, as a main source of misunderstanding in this dilemma. As the first author mentioned, complex bureaucracies may be common in countries high in power distance and uncertainty avoidance. This is especially true for countries where legal, political, and economic infrastructures are not well developed, and personal networking plays a large role in professional relationships and in the attainment of professional and personal goals (Michailova & Worm, 2003). Personal networking may shape decisions on promotions, acceptance into educational programs, and hiring. For example, an individual in a high-power position may be asked to recommend his or her protégés for a job opening, and this recommendation, rather than a hiring committee, would steer the hiring decision.

Personal networking is grounded to a great extent on trust. The choice of whom to trust and to what extent is based on proofs of reliability. If a professor recommends a doctoral student for a fellowship, but the professor does not deliver in the end, the trust relationship may be tainted. Losing personal trust is frustrating because it takes time, effort, and commitment to build trust. In addition, the professor may lose face because, in his or her high-power position, he or she was not able to deliver a fellowship after personally recommending a student.

The best way to avoid the professors feeling insulted due to their nominees not being selected would be to inform the professors up front about the cultural differences in the nominee selection process and let the professor know that his or her recommendation does not totally determine who will be awarded the fellowship. Professors will communicate this information to their protégés to avoid a potential loss of trust and a feeling that their status is being challenged.

REFERENCES

Brislin, R. W. 2008. *Working with cultural differences: Dealing effectively with diversity in the workplace.* Westport, CT: Greenwood.

Gelfand, M. J. 2012. Culture's constraints: International differences in the strength of social norms. *Current Directions in Psychological Science,* 21(6): 420–424.

Hofstede, G. H. 2001. *Culture's consequences: Comparing values, behaviors, institutions, and organizations across nations* (2nd ed.). Thousand Oaks, CA: Sage.

House, R. J., Hanges, P., Javidan, M., Dorfman, P. W., & Gupta, V. 2004. *Culture, leadership, and organizations: The GLOBE study of 62 societies.* Thousand Oaks, CA: Sage.

Michailova, S., & Worm, V. 2003. Personal networking in Russia and China: Blat and guanxi. *European Management Journal,* 21(4): 509–519.

53 Balancing Profession with Ego: The Frailty of Tenure Decisions

P. Christopher Earley

While there are many challenges facing a developing scholar, one of the most salient issues is achieving tenure, or long-term institutional commitment. Of course, tenure is a consequence of being an effective scholar, educator, and colleague. To an individual, tenure provides a sense of security and the opportunity to pursue a research and educational agenda that may diverge from mainstream emphases, and thus to take on new risks and adventures. From an administrator's perspective, such as that of a dean, and the personal perspective I'm presenting in this case, tenure represents a number of commitments made by an institution to a faculty member. Financially, tenure represents an investment in an individual easily amounting to $5.5 million in salary (conservatively based on 20 years of $200,000/year) and an additional $3 million in benefits in today's dollars. Depending on specialization, a business school professor may be expected to earn $8–10 million in salary and benefits during a post-tenure career.

In terms of research, a positive tenure decision commits a faculty slot for a career that will both expand and limit the organization's ability to respond to new market demands. A faculty member whose research remains stagnant and non-evolving commits critical resources of a university over a very long period in ways that might have been otherwise deployed. On the other hand, an innovative and dynamic scholar and educator who commits to an institution can create tremendous success and impact for a university.

What does this mean? A tenure decision must be based on a scholarly decision rooted in the significance and impact of a person's work and not overly simplistic metrics of publication (e.g., the plague of so-called A-list publications), internal politics, or any other irrelevant consideration.

Over the course of my career (including seven faculty positions and three deanships across three continents), I have witnessed the mechanization and politicization of tenure decisions at some universities. In one case, my role was questioning a tenure decision by individuals, who employed simplistic metrics of publication (how many so-called A-list publications that were published on a list were developed in-house), disregarding, in my opinion, evidence of scholarly impact and educational contributions. A university review of the case admonished several input reviews that occurred prior to my review as not accurately representing the actual data from the dossier. I believed my role and obligation as a scholar was to support, or oppose, the tenure candidate based on a full evaluation of the candidate's full record, including published and in-process work, teaching reviews, external evaluations, and service contributions.

Rather than engaging in an open scholarly discussion about the candidate whose tenure was under consideration, several faculty members chose to take my disagreement with their recommendations as an attack on their judgment and, I believe, a threat to their self-identities as academics. After a careful review at various levels of the university, a final tenure decision was made in the case. While the outcome of the candidate's review was consistent with my recommendation, the consequence for me was very negative and led to numerous professional and personal attacks. What I'm not aware of is how this has impacted the individual whose tenure was granted.

In thinking about this incident, and having the luxury of hindsight, I ponder now my decision to support this case. In the end, my decision would remain the same; compromising on professional judgment is unacceptable in a scholarly environment. But be forewarned: personal agenda and ego are very disruptive to our desired academic context. We would like to think that these types of decisions (tenure, grants, promotion, etc.) are based on a careful weighing of evidence and substance. My experience is that even at the very best universities, irrelevant criteria may be introduced into key decisions; however, the temptation of ego and minds "made up in advance of formal review" escalates at increasingly mediocre institutions with individuals trained with a mindset that relies on overly simplistic evaluation schemes.

As a dean (now in my third deanship), my strongest recommendation is for a new scholar to seek the type of academic environment best suited to one's aspirations. Your colleagues' views must be weighed carefully in how they approach their profession. Note the adage conveyed to me by an old friend and tremendously successful scholar and educator: "Never try to teach a pig to sing; it wastes your time and annoys the pig" (attributed to Robert Heinlein in *Time Enough for Love* [1973]).

54 Fidelity and Responsibility in Leadership: What Should We Expect (of Ourselves)?

Donald J. Foss

From the APA Ethical Principles of Psychologists and Code of Conduct:
Principle B: Fidelity and Responsibility
Psychologists establish relationships of trust with those with whom they work. They are aware of their professional and scientific responsibilities to society and to the specific communities in which they work. Psychologists uphold professional standards of conduct, clarify their professional roles and obligations, accept appropriate responsibility for their behavior and seek to manage conflicts of interest that could lead to exploitation or harm.

A faculty member claimed that his academic unit, and in particular myself as its head, did not properly credit him for his peer-reviewed publications. As a result, he asserted, his raises were inappropriately low for a number of years. He requested money to make up for the claimed losses, and a change in base salary to bring it to the level he thought proper. Importantly for our present topic, my colleague said that his beliefs about some nonacademic matters were well known and unpopular in the unit, and that they were the real reason for his low raises. In short, he claimed that the unit was violating its own standards for determining merit increases and that I knowingly abetted this action. If correct, then I was behaving unethically.

My colleague filed a formal grievance in accordance with university procedures. He also hired a lawyer, thereby raising my blood pressure a few points. The university then provided an attorney to assist me, and another to serve as counsel to the three-person hearing committee that was composed of faculty members from other units on campus. The rules of procedure provided for the calling of witnesses, submitting documents and data, and so on. Neither side took issue with the process, which seemed quite fair.

Some background: The unit in question had a faculty committee that reviewed the annual reports provided by each faculty member and then recommended raises consistent with the available money. As head of the unit, I was directed to provide separate and independent advice about the appropriate raises. Both sets were then forwarded to the next higher authority on campus. As came out during the hearing, my raise recommendations for the grievant were generally somewhat higher than those of the budget committee, though lower than one might predict from simply counting the number of his papers published in refereed journals.

The gist of my defense – made both on my behalf and on behalf of the unit – was that the grievant's publications were scattered in topic. To put it somewhat tendentiously, they seemed to constitute a collection of papers more than a program of work. I further noted that the faculty, working through their budget committee, valued programmatic work that made a difference to a subdiscipline within the field. While frequent publishing was a plus, the grievant's record was not considered to be as strong as those from faculty members who made more programmatic contributions.

In response, my colleague said that the need for programmatic work was not stated in the unit's by-laws and that I had not explicitly conveyed this criterion to him. I believe he considered the "non-programmatic" argument to be a smokescreen, one thrown up to hide the alleged bias against him. It was clear to me then, as it still is now, that he believed what he was saying; his grievance was not a ruse.

The witnesses told the hearing panel that the grievant was a long-time member of the faculty, that he was himself a highly experienced member of the budget committee, and that the preference for programmatic work was part of the unit's well-known culture. Further, one of the witnesses said that he was not even aware of the grievant's "outside" beliefs and thought that many others were similarly in the dark.

Toward the end of the hearing the members of the panel asked us all a number of questions and politely excused us. Then, as faculty members will do, they deliberated. A few weeks passed during which I fretted – the prospect of being found unethical by my colleagues was not a welcome thought. At last the decision letter arrived.

The hearing panel unanimously found against the grievant. Indeed, in their report they called my handling of the matter "exemplary and commendable."

In retrospect, while I still believe I acted ethically, I doubt that my behavior was exemplary and commendable. Speaking somewhat loosely, I'm not sure I occupied that space defined by the intersection of real psychology, sound leadership, institutional responsibility, and ethics.

It wasn't surprising that my colleague was disappointed by his raises, given his publication record. He was not a slacker. Of course, plenty of faculty members are disappointed by their salary increases each year, and some express it to the head of the unit, occasionally with passion; that comes with the job. But his case was not the run-of-the-mill kind, and I should have recognized it. More importantly, I should have proactively called him in and given him the bad news and my explanation for it face to face. Thinking back on it, I believed then that his attributions would square with those of the faculty committee (and mine) because he had been in the unit for a long time and was a savvy person. But, of course, a negative result invites alternative attributions, and from his point of view he had a justifiable one.

After 25 years as an academic administrator, I could – unfortunately – multiply this example. I believe that it and many of the others had a common source. To wit, at that time our university had not properly advised and trained its leaders in how to give direct and honest feedback. I further believe that this is still true in many places, even those with faculty unions. Given the fact that academic leaders usually come from the faculty and that most of them expect to return to a full-time faculty role, many find it highly uncomfortable to deliver bad news to their colleagues. They prefer to avoid the conflict that such messages can induce. But such avoidance behavior can, and often does, yield even deeper conflicts, higher blood pressure, and less desirable outcomes than would a forthright and even uncomfortable initial discussion. I think it appropriate to ask the question whether such behavior is consistent with the requirement for psychologists to "uphold professional standards of conduct."

My institution thoroughly supported me when I "got in trouble" – it had in place a fair grievance process and provided me with a very smart lawyer and lots of her time. But importantly, the university did not support me (and others in similar positions) in a more fundamental way – by training us to provide regular and honest feedback to others, among other things.

Not only that. In fact, the person to whom I reported did a poor job of reviewing my performance – the useful, honest feedback provided to me asymptotically approached zero. Higher-level administrators were, with an occasional exception, not modeling the right behaviors. Fidelity and responsibility in leadership requires more than "being right" in a technical or legalistic sense. That's not enough; we should expect more of each other and, importantly, of ourselves.

I believe the academy is still subject to analogous, knotty problems. Solving them requires that senior administrators recognize the need to provide appropriate training to new leaders and more than pro forma feedback on their performance.[1] Furthermore, those who are asked to take on positions of responsibility should request – even demand – such training and feedback as a condition of stepping into the job. When that happens, then the level of "trust with those with whom they work" will have a much-improved chance of staying high.

NOTE

1 I'm pleased to report that I got the opportunity to institute such training programs later in my administrative career.

55 To Thine Own Self Be True

David Trafimow

Tenured professors have a great deal of power over the promotion and tenure prospects for assistant professors that are on a tenure track but not yet tenured. The sad fact of the matter is that tenured professors can cause assistant professors to be denied tenure, based on the slightest of whims even in the face of adequate, or even better than adequate, performance on the part of the assistant professors. If an assistant professor angers a tenured professor, it is at a very real risk of eventually having to leave the university. Any assistant professor that has kept even one ear open has heard relevant horror stories, and consequently there is a strong perception of pressure to get along well with everyone. And yet, might there be situations where ethical considerations come into play that may put pressure on an assistant professor to take the risk of offending tenured professors?

Early in my career I encountered a situation of this type. I was a new assistant professor (one tenured professor kiddingly referred to me as "the new kid on the block"), and I naturally desired to get along well with everyone. As luck would have it, I arrived just in time for an external evaluation of the department, which comprised four areas: clinical psychology, industrial-organizational (I-O) psychology, developmental psychology, and applied-experimental (A-E) psychology (please do not ask me what this last one means!). As a social psychologist that did not fit well into one of the first three areas, I was assigned to the A-E area. The results of the general evaluation were predictable. The clinical psychology area was deemed to be one of the best in the country, the I-O area also received rave reviews, the developmental area received some sympathy based on the excellence of its individual members in the face of there not being enough of them, whereas the A-E area was criticized in no uncertain manner. Not surprisingly, the negative evaluation of the A-E area was not received well by the members of that area (except for me; I agreed with the review), and the general consensus was to write a protest letter.

So my ethical dilemma was before me. It was clear to me that the negative external evaluation of my A-E area was valid, and so my signing the protest letter would be a misrepresentation of what I actually believed. Put more plainly, my signing the letter would constitute a lie. So, on the one hand, I could act according to my conscience and not sign the protest letter. This would have the advantage of being the honorable thing to do. But it would have the disadvantage of incurring the wrath of the tenured professors in the A-E area, with the substantial likelihood of that wrath being expressed later in the form of poor evaluations and my being denied eventual tenure. On the other hand, I could hold my nose and sign the letter. This arguably was the rational thing to do. The tenured professors in the A-E area would be appeased and the tenured professors in the other areas would assume, as was pointed out to me, that I had been strong-armed into signing. Thus, nobody would blame me, and my tendency to accrue publications at a fast rate would ensure that I would gain eventual tenure and promotion.

So what did I do? As I just stated, a rational expectancy-value calculation based on the obvious risks and benefits, and probabilities thereto appertaining, dictated that I should sign the letter. Worse yet, I knew well that this was so. Yet, I could not do it. To this day, I cannot generate a decisive argument in favor of my decision to not sign the letter. Fortunately, I did not suffer for my irrational behavior. The tenured professors who were not in the A-E area interpreted my not signing the letter as indicating high moral character and courage, and so my moral status actually increased in their eyes. Of course, this was a very generous interpretation, and it might be more plausible that I simply was lacking in rationality. Or as the line goes in New Mexico, perhaps I was "one taco short of a combination plate!" In addition, I later took a job at New Mexico State University in Las Cruces, where I currently reside, and my wife and I enjoy living here. So I did not suffer, in any way, for my irrationality. But I sometimes look back and wonder about two things. First, how could I have been so irrational? Second, what would I do now in a similar situation?

In considering how I could have been so irrational, it is possible to ask whether I really was irrational, or whether my behavior could be interpreted as rational if additional factors are considered. For example, we might widen the field by considering rationality in terms of what is best for most people. From this point of view, it seems obvious that society, in general, suffers if faculty members lie routinely just because they think it is in their best interest or because they have been intimidated into it. Taking the General Interest seriously as being more important than my own Specific Interest implies that perhaps my behavior actually was rational. In addition, even from the point of view of my Specific Interest, do I really want to be the kind of person who will tell lies out of personal fear? What kinds of self-deceptions would I have to practice in the future to prevent me from despising myself for my cowardice? What kinds of relationships with others can I have when they figure out that I am the kind of person who will tell lies under what really is only moderate provocation? After all, although there was the threat of not getting tenure, and this certainly is not trivial, the extent of the threat fell well short of the extent of the threat that there would have been had I been faced with death by torture!

Now that I am a lot older, and have had the opportunity to study the lives of others over an extended period of time, my experience has led me to a conclusion that I cannot back up with any formal data, but that I believe anyhow. When people make moral compromises, it does something to them. They become the kinds of people who make moral compromises! The first moral compromise makes the next one easier, and so on. In turn, being the kind of person who makes moral compromises decreases people's abilities to have meaningful, close, long-term relationships. Although a person who cannot be trusted when "the rubber hits the road" might be kind, friendly, and fun to hang around, would you really want to have this person as your best friend? Would you want to be married to this person? So don't be that person!

I would summarize as follows. From a restricted point of view of costs and benefits, my behavior of defying the tenured professors in my area was irrational. However, it arguably becomes rational from a less restricted point of view, and particularly if the General Interest is considered.

First General Principle. Try to avoid being dishonest, even in socially excusable ways, and even if you perceive pressure from your colleagues.

Second General Principle. When calculating expected value, include the subtle and intangible considerations in your calculations, as well as the obvious and tangible ones.

Third General Principle: If the Second General Principle involves calculations that are too complex, obey the Shakespearean maxim: To thine own self be true.

56 When Things Go Bad

Robert J. Vallerand

Context

Over the past 25 years or so, I have had the pleasure of serving as president of various associations, including the Quebec Society for Research in Psychology, the Canadian Psychological Association, and the International Positive Psychology Association. Serving as president can be rewarding, as one gets to propose and institute changes that will bring forward the association's objectives and activities. At the same time, some problematic situations may arise that merit concrete action. As the lead officer, it typically falls on the president to engage in appropriate action in these situations so as to find a solution to such problems.

The Problem

While I was serving as president of one of the aforementioned associations, the following situation took place. An individual had submitted two abstracts for two symposia (one presentation in each of the two symposia) for the association conference. Both were accepted. However, when the time came to present the information at the conference, the person did not present the content that he or she had submitted for approval. In the first symposium, he or she presented instead some materials whose content was shocking to some of the audience. The individual explained his or her decision to present a different content by simply indicating that had he or she submitted an abstract on the actual content of the talk, it would have never been accepted. In the second talk, instead of presenting the proposed talk, the individual used the time allocated to the talk to engage in some bashing of the association.

Clearly, such behavior is unacceptable on at least four counts, as it entails: (1) presenting unannounced potentially shocking content to a captive audience; (2) failing to present material that the person had agreed to present; (3) presenting content that was not reviewed by the scientific committee; and (4) using conference time inappropriately, as there are other venues for voicing one's discontent. Unsurprisingly, several members of the association were offended and complained to members of the executive committee, including myself. Action was to be taken.

Action

Here is what was done in this situation. First, the issue was thoroughly discussed at the next executive committee meeting, and a course of action was decided, which I implemented. First, I contacted presidents or past presidents of other associations to see how their association would handle that situation. These individuals were very gracious with their time and made insightful suggestions that ranged from ignoring the situation to kicking the person out of the association. Second, with this information in hand, I contacted again the members of the executive committee, and a middle-of-the-road approach was chosen. Specifically, it was decided that I would talk to this individual and formulate a warning. Third, I did talk to the person. I first asked for his or her side of the story to see if what we had heard was true. When the story was confirmed, I clearly indicated that such behavior was inappropriate on several counts (see the list presented earlier) and asked the person if he or she agreed with my assessment. The person agreed and apologized. Then, I proceeded to mention that if this would ever occur again, I (or future officers of the association) would have no choice but to bar him or her from entry into both the association and future conferences of the association. The person was very cooperative and indicated that he or she would not engage in such behavior again. As far as I know, he or she never did. Finally, I reported back to the executive committee, whose members seemed satisfied with the course of action and its conclusion.

Conclusion

When people behave unethically within the context of an association, there are a number of courses of action that can be taken. Such action may depend on a number of factors including, among others, the gravity of the behavior and the settings in which it took place. In the situation described in this chapter, the course of action included asking officers from other associations for advice as well as directly talking with the perpetrator. I believe that these two steps were highly influential in bringing about a fruitful conclusion to this problematic situation. Of course, more severe courses of action could also have been used (such as kicking the person out of the association) and might have been effective in this very situation. However, I tend to believe that we are in the business of psychology, and as such we should try to use it whenever possible to bring about a resolution of a problematic situation without being unduly harsh on the culprit.

57 Commentary to Part IX

Susan T. Fiske

Culture, loyalty, and ego each contribute to personnel decisions in academia, maybe even more so than in business, because our criteria rarely count dollars and widgets. Assessing intellectual merit is difficult and subjective, paving the way for bias or valid differences of opinion.

Culture entails not only distinct national and regional norms, as described in contributions to this Part, but also local university, college, and discipline norms. Both deciders and candidates would do well to consult wise heads about the applicable expectations: standards, precedent, process, transparency, feedback, accountability, recourse, and even etiquette.

Loyalty involves who argues for whom, group agendas, shifting alliances, power structures, and other aspects of local politics. These are harder to discover except by observation, but mentoring advice may help.

Ego issues in personnel cases involve both deciders and candidates. People want their voices heard in the process, on both sides. When the procedures are fair, people are more likely to accept even undesired outcomes as fair.

Part X

Reviewing and Editing

58 The Ethics of Repeat Reviewing of Journal Manuscripts

Susan T. Fiske

In my capacity as a journal's associate editor, I received a reviewer's cover note, which reported having reviewed the paper for another journal, having read the new submission, and finding the manuscript little changed in response. This is not uncommon, in my experience as a journal editor. Sometimes reviewers refuse to review the same manuscript twice, to prevent putting the author into double jeopardy, potentially putting themselves into the position of vetoing a paper. Sometimes reviewers ask the editor's advice about whether to re-review. This itself is an ethical dilemma.

After considering this, then deciding also on the basis of two other reviews, one positive but critical and one negative, as well as my own independent response – which was to remain unconvinced by the paper – I rejected it. The author wrote politely to question whether the repeat reviewer had acted ethically, stating that the review was reportedly identical to the previous review. The author did raise the issue of double jeopardy and the reviewer reportedly submitting verbatim reviews twice, despite the author reportedly having substantially revised the paper. The author did not, however, request a reconsideration of the editorial decision.

I responded, thanking the author for thoughtful consideration of the issues involved. I communicated that the reviewer did indicate in a cover note to me having reviewed the manuscript already for another journal, but felt that the manuscript was not much revised from the earlier version. I considered this information in the context of the other two reviews and my own reaction, so the configuration guided the outcome. That person did not unduly affect the decision, which would have been the same without that reviewer. As a social cognition researcher, I do know that decision makers cannot know what influences their responses, but in this case, even if I had had only the two negative responses (mine and the new reviewer's), that would have prevented the paper's acceptance.

Although others may disagree, I did not feel it was unethical on the reviewer's part, although if the person were asked yet again, the person should not keep reviewing the paper because then a line would have been crossed, in my opinion.

This case illustrates for me how differently two parties (repeat reviewer and frustrated author) can interpret the same transaction (respectively perceiving an unresponsive, unrevised manuscript versus a substantially revised one) and how much where you sit determines where you stand on such issues.

59 Bias in the Review Process

Joan G. Miller

This example concerns the importance of not being swayed by political or other self-interested considerations in evaluating the merit of scientific research and the need to remain open to theoretical perspectives that may be challenging of dominant paradigms. The example involves specifically the downplaying of critical opinions that I observed as a member of a site visit panel charged with appraising the merit of a grant application exploring neurological bases of psychology. As a member of this site visit panel, I observed practices, described in this chapter, that were designed to shield the grant application from negative appraisal.

One of the practices involved the site visit review panel dismissing dissenting opinions that raised questions about certain basic assumptions of the paradigm involved. I had been selected for the site visit panel as an expert on issues of culture who had been brought in to provide a perspective complementing that of the other faculty on the panel, whose expertise, if not disciplinary affiliation, was in biology and neuroscience. As members of the review panel, we were asked to write individual critical reviews of the funding proposal, with these reviews subject to discussion and further input by others in our group. In my review, I articulated some of the same critiques of neuroscience that have been raised in recent years by major theorists who, while highly supportive of neuroscience work in psychology, have raised concerns about its tendencies, in cases, to adopt stances that are reductionist or deterministic, and who have pointed to the difficulties entailed in mapping brain processes onto constructs in psychology (e.g., Barrett, 2009; Kagan, 2007). The other faculty on the site visit panel objected to my raising any of these types of concerns and asked to have everything that I had written deleted from the final site visit report (something that had not been done in the case of comments written by anyone else on our panel). I was given little opportunity to defend my views and no opportunity to include my concerns in the final report as a dissenting opinion.

On this same panel, I observed as well an embellishment of our site visit report that occurred in appraising the broader impacts section of the research proposal. The faculty on our site visit team completed the broader impacts section of the site visit report by a process that involved cutting and pasting text from the broader impacts criteria specified on the funding agency website, and including this pasted information in the site visit report along with a statement, in each case, that the present grant application fully met each criterion. In doing this, our site visit panel failed to take into account any specifics about the proposal under consideration and ended up with a broader impacts section of the final report that focused exclusively on positive information, with no mention of any limitations at all in this domain.

If I were in the same situation again, I would have taken steps to communicate more fully my misgivings about our grant review process to the funding agency official who had been present during the deliberations. I also would have requested that my name be withdrawn from the list of panelists who were part of submitting the final site visit report in order to communicate my dissent from what I considered to have been the somewhat biased review process.

A general principle this example illustrates is the need to avoid favoritism in evaluating the merit of scientific research and to be open to, rather than silencing, dissenting viewpoints. The example I narrated occurred several years ago during the period when interest in neuroscience in psychology was just beginning to undergo an exponential increase. The members of our site review panel seemed motivated by a recognition that not only the success of this specific grant application but also the success of the reorganization then under way by federal funding agencies like NIMH to give priority to neuroscience research could be put in jeopardy by open discussion of the challenges entailed in neuroscience being able to fulfill its promise of revolutionizing the discipline. However, one might argue that the pointed criticisms that have been made of at least some early work in neuroscience (e.g., see also Vul, Harris, Winkelman, & Pashler, 2009) might have been less needed had work in this tradition remained more open to major conceptual and empirical challenges to the paradigm. More generally, it must be recognized that only by being receptive to dissenting opinions and to critical voices can the self-corrective processes and integrity of science be maintained.

REFERENCES

Barrett, L. F. (2009). The future of psychology: Connecting mind to brain. *Perspectives on Psychological Science, 4*, 326–339.

Kagan, J. (2007). A trio of concerns. *Perspectives on Psychological Science*, 2, 361–376.

Vul, E., Harris, C., Winkelman, P., & Pashler, H. (2009). Puzzlingly high correlations in fMRI studies of emotion, personality, and social cognition. *Perspectives on Psychological Science*, 4, 274–290.

60 The Rind et al. Affair: Later Reflections

Kenneth J. Sher

In the course of one's professional life, you might step on a landmine and have no idea that you have, until it explodes a long time after you've traversed the minefield. Such an event happened to me personally about 15 years ago when I was serving as associate editor at *Psychological Bulletin* and accepted for publication a manuscript reporting a meta-analysis on the long-term effects of child sexual abuse (Rind, Tromovitch, & Bauserman, 1998). Although it appeared in print for several months with little controversy, a firestorm of criticism ultimately arose, with the flames fanned by controversial radio personalities (e.g., Laura Schlesinger), Christian conservative political groups (e.g., Family Research Council), members of Congress, and pro-pedophilia activist groups (e.g., North American Man/Boy Love Association), among others. The paper was ultimately condemned, with no dissenting votes, by both houses of the U.S. Congress and presented a major political and public relations challenge to the American Psychological Association (APA). Many of the details are well known and, indeed, I (Sher & Eisenberg, 2002) and many others (see special issue of the *American Psychologist*, 2002, especially Lilienfeld, 2002) have commented on the now infamous "Rind et al. affair" in great detail. These and other published commentaries highlight numerous issues concerning the nature of peer review, the relationship between science and values, and the politicization of science.

During the review process itself, which spanned multiple review cycles, I did not perceive any ethical challenge. Although clearly an oversimplified view of my editorial role, I tended to see my job as "calling balls and strikes." I made such calls based on the best available information available to me, primarily the informed opinions of the reviewers and my own independent reading and evaluation of the manuscript. This is not to say that I was then or am now 100% reliable in my decision making; no umpire or editor is, and it's often unclear if a given pitch is inside or outside the strike zone. In addition to our own nonsystematic error that leads to variation in decision making, we all have systematic biases of which we may only be aware to a limited extent. Additionally, it is clear that thresholds for publication vary not only across but also within journals as a function of the relative balance between the number and length of submissions and allotted journal pages as well as the relative balance of content published in the journal. Moreover, editorial decisions concerning review articles of the type *Psychological Bulletin* published may be particularly vulnerable to various biases because there is less variability in method (especially a well conducted meta-analysis), and a key issue concerns the potential impact to the broader discipline, a judgment that is inherently speculative to varying degrees.

Although not feeling ethically challenged at the time, I was clearly aware on some level that there were political and legal implications of the paper, which was reflected in my insisting that the authors distinguish legal and scientific concerns and not conflate the two, as well as explicitly acknowledge that socio-legal definitions should not be changed based on these or similar findings. But none of the reviewers raised any concerns of potential underlying controversies. Indeed, if anything, the sense I got from the reviewers was that if there was an issue that weighed against publication with respect to the empirical findings, it was that they were somewhat pedestrian. That is, the reviewers didn't raise any red flags of potential controversy.

Of course, the question arises: Who are these reviewers? Would another group of reviewers have had greater concerns? Indeed, this is always a fundamental question underlying the integrity of peer review. The editor must try to identify an adequate number of potential reviewers with the relevant expertise in the area and a broad appreciation of the larger field to evaluate fairly the totality of the manuscript's contribution. I felt I had done a reasonable job of that, although obtaining reviews from the most appropriate reviewers is not something over which the editor has total control. Indeed, as is often the case, not all reviewers I approached about contributing a review were able to provide one. I mention this only because I subsequently heard a criticism, leveled against the peer review process that led to publication, that "Expert A," a leading figure in this area, was not one of the reviewers, with the implication that his/her opinion was deliberately avoided or inadvertently overlooked, and that had this reviewer's opinion been obtained, the manuscript surely would have been rejected. Such information was presumably obtained directly or indirectly from "Expert A" and not from the journal, owing to the policy of keeping reviewers' identities confidential. In fact, I had initially requested a review from "Expert A," who turned down the opportunity to review.

The first ethical challenge I personally experienced in this case was the conflict between maintaining confidentiality of review and providing assurances that there was no attempt to avoid specific appropriate reviewers and that, in fact, this well-known expert was approached about reviewing and declined the opportunity. Of course, I had no choice but to maintain confidentiality of review but would have welcomed the opportunity to correct the incorrect perception that there was a lack of due diligence in obtaining the most appropriate reviewers.

Ultimately, the APA launched its own investigation of the decision-making process, and my own editorial work was as scrutinized as any that I've ever heard of. APA also reached out to the American Association for the Advancement of Science (AAAS) to conduct an independent evaluation. While such scrutiny of one's work by outside groups is not something one seeks out, over the course of my career I've come to the conclusion that transparency is the general principle that prevents and corrects most of the problems we're likely to face in research and scholarship. In reviewing all of the editorial correspondence that was being pored over by others, I took comfort in the fact that I felt I had done a reasonable job vetting the manuscript's issues, weighing the reviewers' comments, using my own judgment, and communicating clearly with the authors. While I had no idea as to whether someone else would have made the same decisions I had made, I did feel that the integrity of the process was clearly documented. Apparently, both the APA and AAAS found nothing over which to question the editorial process that led to my decision to accept the manuscript. But it is certainly possible they could have, or another group could have arrived at a different conclusion.

Fortunately, most scientists serving in editorial capacities never find their editorial decisions leading to national debate and resolutions of condemnation on the House and Senate floors. However, to varying extents, the factors that led to this state of affairs are likely quite common, and it is probably helpful to identify what these are in order to prevent (or minimize) editorial decisions from becoming overly politicized, with powerful groups and individuals attempting to adjudicate what is good science on the basis nonscientific values, partisan political interests, and personal biases. Indeed, with increasing concerns over questionable research practices (Martinson, Petersen, & de Vries, 2005), scientific misconduct (OSTP, 2005), and poor replicability (Giles, 2006), the challenges faced by the editor are now in bolder relief than in the past, because the perceived need to reform our science is currently great, and as the gatekeeper into the scientific literature, the editor plays an increasing pivotal role.

So, what can be done to prevent such problems in the future? I don't have an exhaustive list, but the following seems to be a reasonable set of suggestions to consider.

1. *Identify and successfully recruit the most appropriate qualified reviewers.* While this may seem obvious, this may be more challenging than it seems on the surface. Identifying optimal reviewers requires a diligent process that overcomes normal biases (e.g., availability heuristic) and involves some informal vetting of appropriate experts. This is becoming increasingly difficult, or at least it seems so to me, since some colleagues are reluctant to take on too many reviews, and a stellar list of invitees doesn't always translate into an equally stellar list of obtained reviews. The issue of how to engage busy researchers in reviewing more manuscripts goes beyond this short essay but must be tackled at many levels, including stronger socialization of the values of the peer-review system, more of a personal touch in soliciting reviews, and possibly reward contingencies for meritorious review service.
2. *Review the manuscript for potential "hot button" issues even if they aren't raised by reviewers or the authors.* Indeed, many of us are "tone deaf" when it comes to content that might irritate or inflame some constituencies of our research. The goal of greater sensitivity to these issues is not to find cause to reject a paper, but rather to take steps to ensure the research is presented in a way that is less likely to be mischaracterized or misrepresented. Such steps could range from judicious editing to minimize misinterpretations to explicit discussions of what the research is saying and not saying, and the boundary conditions of the empirical phenomena and their generalizations to critical applications.
3. *Provide opportunities for published commentaries from likely dissenting experts.* This approach allows for many "sins" to be addressed in that it provides a more extensive vetting of issues than is possible through the normal publication process. In my opinion, the original authors should always be given the opportunity for rejoinders, given that the issues raised in commentaries are often targeting specific aspects of the original authors' work that may need further elaboration or readers to fully evaluate any emerging debate. While comments on previously published work should be an option available to commentators, publishing commentaries and rejoinders along with the target article could do much to prevent unnecessary controversies to occur outside of scholarly venues.
4. *Provide opportunities for (moderated but not refereed) readers' comments that are public and provide an ongoing forum for issues.* The digital revolution has provided technologies enabling readers to provide commentaries on publishers' websites, which allows for much broader debate in a well-defined venue than has been possible in the past. To the extent that many of the problems associated with the Rind et al.

affair stemmed not so much from the debate per se but from the settings in which the debates occurred, fostering more helpful debate venues could focus energies on more constructive dialogue.

The social and behavioral sciences have met with increasing criticism from various groups outside the academy because of scientific research that presents ideological challenges to entrenched political interests. Maintaining the integrity of peer review and fostering informed debate within (and associated with) scholarly venues seems critical for improving the quality of debate and reducing opportunities for mischaracterizations. While it would be naïve to suggest that academics can inoculate their ideas from political mischaracterizations and crass demagoguery, we should be aware that we can probably do more to minimize the possibility that this will occur, given that the consequences can be very damaging to the public perception of our fields and public funding for our efforts.

REFERENCES

Giles, J. (2006). The trouble with replication. *Nature*, 442, 344–347.

Lilienfeld, S. O. (2002). When worlds collide: Social science, politics, and the Rind et al. (1998) child sexual abuse meta-analysis. *American Psychologist*, 57(12), 176–188.

Martinson, B. C., Petersen, M. S., & deVries, R. (2005). Commentary: Scientists behaving badly. *Nature*, 435, 737–738.

OSTP. (2005). *Federal Policy on Research Misconduct*. Retrieved from http://www.ostp.gov/html/001207_3.html

Rind, B., Tromovitch, P., & Bauserman, R. (1998). A meta-analytic examination of assumed properties of child sexual abuse using college samples. *Psychological Bulletin*, 124(1), 22–53.

Sher, K. J., & Eisenberg, N. (2002). Publication of Rind et al. (1998): The editors' perspective. *American Psychologist*, 57(3), 206–210.

61 Me, Myself, and a Third Party

Steven K. Shevell

Background

Most scientific journals require that a manuscript submitted for publication not be under consideration for publication elsewhere. Another tenet of editorial processing is that authors submit their paper as a confidential communication, which generally means the editor and reviewers may not disclose the content of the paper to a third party. But what if the third party is the editor himself?

This happened to me as an editor at journal X. I asked two experts to review a submitted manuscript; both suggested significant changes. The reviews were communicated to the authors, who were invited to prepare a revision. Subsequently, an author contacted me more than once to clarify specific aspects of a (presumed) revision. All fine so far. I anticipated seeing the revised paper shortly.

The manuscript reached me soon afterward, but not by the route I expected. It was accompanied by a request to review the paper for a different journal (call it Y).

Two questions immediately came to mind: (1) how to respond to the editor of journal Y and (2) what to do (if anything) about the submission still pending at journal X. The options for (2) raised the issue of whether I, as a potential reviewer for journal Y, could reveal receiving this paper to myself as editor of journal X (a third party from Y's perspective).

Points for Consideration

1. When an editor invites authors to make revisions, the manuscript remains under active consideration. Journal X as policy sets a specific deadline for receiving a revision. This is communicated to authors together with the reviews. (Of course, an author may withdraw a paper at any time, but this was not a factor here.) If the deadline for revision had passed, the authors might reasonably have assumed the paper was withdrawn. The deadline, however, was in the future.

In the eyes of journal X, the paper was still under consideration for publication when the authors submitted it to journal Y.
2. As editor at journal X, I was bound by the ethical standards, including confidentiality, of X. As a potential reviewer for journal Y, my obligation in a technical sense was vague because I had not agreed to do a review. The request to review from Y was unsolicited. Nonetheless, I assumed I was bound by the confidentiality standards at journal Y.
3. A pledge of "confidentiality" is made to authors and requested of reviewers, but confidentiality is often poorly defined or not defined at all. Confidentiality generally prohibits disclosing the contents of a manuscript to a third party.[1] Some journals have a broader definition that prohibits disclosing other information about a submission, including its status in the editorial process or even whether it was received.[2]
4. Journal X clearly prohibits any other journal from simultaneously considering a paper for publication. Given that an author contacted the editor of X (me) about revisions, there was reason to believe the authors thought the paper was still under consideration at journal X. On the other hand, the authors may not have been clear about the definition of a paper "under consideration"; no version of the manuscript was in the hands of reviewers at journal X when the paper was submitted to journal Y.

Action Taken

(1) I declined the request to review from journal Y as a conflict of interest. (2) As editor of journal X, I withdrew the paper from further consideration at journal X, informing the authors that I discovered the manuscript was under consideration for publication elsewhere.

Reasoning

The justification for declining the review at journal Y was bias because I already had formed an opinion of the paper during the review process at X. This was clear-cut.

The reason for withdrawing the paper from further consideration at journal X was to maintain the policy at X that a paper under consideration may not be submitted to another journal. The manuscript obviously had been submitted elsewhere. If there was a gray area, it was that I learned of the submission to Y in the role of a potential reviewer there. Was enforcing the policy of journal X against dual submission a violation of confidentiality at journal Y? At the time, I reasoned it was not because

I considered confidentiality to extend to revealing only scientific content of a manuscript, not to knowledge whether a paper had been submitted for publication.

Aftermath, Afterthoughts, and a Policy Recommendation

Sometime later, one of the authors asserted that I acted improperly by withdrawing the paper from journal X. No reason was given. The editorial decision to withdraw it from X was not changed.

Working down a decision tree, if the authors already had decided to withdraw from X when they submitted to Y (perhaps based on information from our exchanges after they received reviews?), then my action to withdraw the paper from journal X only formalized a decision the authors had made. If, on the other hand, their intent was to try for publication in journal Y and, if unsuccessful, come back to X with a revision, then this was contrary to the policy of journal X, so my editorial action to withdraw the paper was appropriate.

This case highlights the importance of describing fully and precisely what a journal means by confidentiality during editorial processing. A nonsystematic check reveals journals that require it without specifically saying what confidentiality requires. A single definition may not fit all journals, so editorial boards need to adopt a confidentiality policy tailored to their requirements.

NOTES

1 See peer-review publication policies of *Nature*.
2 See publishing ethics policies of the *Journal of Experimental Biology*. For the record, neither of the footnoted journals is X or Y.

62 Commentary to Part X

Susan T. Fiske

Editorial and review processes are open to all the ethical weaknesses of human nature: favoritism, bias, controversy, misrepresentation, and motivated disagreement. Nevertheless, peer review is our best option for endorsing scientific merit and allocating prestigious journal space. The protections of blind – preferably double-blind – review are many. We are less likely to be influenced by loyalty and politics. We have more credibility with the public. We can hope the process is fairer than publishing everything, or the decision of a single gatekeeper, or a quid pro quo mutual backscratching system. Peer review actually does well, assessed against alternative standards.

Part XI

Science for Hire and Conflict of Interest

63 The Power of Industry (Money) in Influencing Science

K. D. Brownell

An issue generating considerable controversy is whether scientists can remain objective and unbiased when accepting money from industry. Large amounts of money change hands between industry and scientists, and this practice has been challenged in prominent articles both in the profession and the press.

Many thought leaders, in fields where industry has a financial stake, have been approached and offered financial benefits. This takes a number of forms including research support, speaking fees, funds to serve on advisory boards, or trips to participate in meetings with other professionals.

The money is consequential, sometimes amounting to more than an individual makes in his or her university salary. This leaves scientists in the difficult position of accepting the enticements industry offers and feeling as if they can help change business practices from within, versus addressing questions of conflicts of interest and fearing the appearance and perhaps the reality of being tainted. The most extreme example in my own experience was an offer of $50,000 from a major food company for less than one day of consulting.

One must guess at the motives of industry. There may be a genuine interest in input from scientists or there may be motives beyond the specific input. In the case where I was offered the payment of $50,000, I agreed to advise the company under three conditions: that I not receive payment, that I cover my own travel costs, and that my name not be listed in any publicity generated by the company. The offer to advise the company was withdrawn.

Critics of industry funding believe it produces a number of negative consequences. The view is that industry cleverly invests in scientists, paying out trivial amounts from the corporate perspective, while receiving much more in return. Research shows that individuals funded by industry tend to produce studies that favor the industries' positions. Such researchers also tend to issue public statements favorable to the industry.

Blind Spots and Self-Forgiveness

I am not the only one to have faced such a challenge. Universities, professional societies, and individuals who take money from industry have what seems to be a serious and predictable blind spot. While they might feel that others are tainted by money, they feel themselves to be immune from influence: They, perhaps uniquely among all others, can remain objective. A clear body of evidence suggests otherwise. In some ways the problem is akin to advertising, where individuals feel others are influenced but they are not.

An example is the contribution of $10 million to the Children's Hospital of Philadelphia from the American Beverage Association (ABA) in 2012. The ABA is a trade association representing the makers of popular beverages, most notably Coca-Cola and PepsiCo.

Prior to this gift being made, the mayor of Philadelphia had introduced a proposal for a tax on sugar-sweetened beverages. After considerable lobbying and public-relations expenses to defeat the tax, the vote was expected to be close. In the 11th hour before the City Council vote, a wealthy Pepsi bottler offered $10 million to the City Council in exchange for voting against the tax. This offer was exposed by the press and withdrawn. After the tax was narrowly defeated and the mayor later expressed his intent to reintroduce the measure, the beverage association gave the $10 million gift to the Children's Hospital.

The Children's Hospital defended accepting the gift, saying they had created firewalls to keep industry from influencing research done at the hospital. In my belief, this statement utterly missed the point. The money effectively silenced the hospital from speaking out against taxes, and bought goodwill in the press and the community at a critical period when the taxes were being reconsidered. I suspect the industry cared little about the research or firewalls but simply wanted silence. The hospital profited, forgave itself for taking the money, and in so doing may have helped industry defeat a measure that would have produced much more than $10 million in public-health benefits. By accepting the gift, the hospital may have contributed to obesity in the community it serves.

Who Will Police This and How?

Individuals in the field do not seem to be in a position to police their own behavior, given how much money there is to be made and the blind spot mentioned earlier. It is routine, in fields in which an industry is affected by scientific research, for top leaders to accept money from the industry.

Perhaps the institutions that employ the scientists might monitor and address the conflicts, but universities employing such scientists often accept money from the same corporate players. Professional associations and societies might be another line of defense, but these groups also often benefit from corporate donations. Examples would be the Academy of Nutrition and Dietetics, formerly known as the American Dietetic Association, and The Obesity Society.

The Fallacy of Disclosure

Professional journals, professional associations, and universities have reacted to conflict-of-interest issues by requiring disclosure of potential conflicts. There are two flaws in this approach, one intuitive and one not. By requiring individuals to disclose potential conflicts of interest, there is implicit or explicit recognition that such conflicts may create problems. Instead of stopping the problems, the solution has simply been to disclose them.

Second, research shows that disclosure may have counterproductive effects. Daylian Cain's research has shown that an audience hearing from a person who discloses a conflict of interest will discount the credibility of the speaker, but only slightly. The speaker, on the other hand, feels licensed to communicate his or her message more strongly, and hence disclosure has the paradoxical effect of enhancing the impact of the person in conflict.

64 The Impact of Personal Expectations and Biases in Preparing Expert Testimony

Ray Bull

A considerable number of years ago I began to be asked to provide "expert witness" reports in relation to a number of court cases. Having agreed to do this I not only (and obviously) chose to read all that I could find on how best to do this; I also read information available to lawyers about how they should "deal with" expert witnesses, so as to prepare myself for dealing with lawyers (who can be demanding, sometimes arrogant, people).

Subsequent to the publication in 1992 by the government in England and Wales of the "Memorandum of Good Practice on Video Recorded Interviews with Child Witnesses for Criminal Proceedings" (that I and a law professor were commissioned to draft), I have been asked on numerous occasions to provide an "expert" report on how well (alleged) child victims/witnesses were interviewed (these having been routinely video-recorded since 1992 and often used as evidence-in-chief).

When asked to provide such reports (and indeed on related topics such as the interviewing of suspects, routinely audiotaped in England and Wales since 1986), I have always insisted that I will not discuss with the retaining lawyers (or any other lawyers) the contents of my report until after I have sent the report in writing to them. Once the report is in, I am willing to change some of my wording only for the purposes of clarity, but I am not willing to change my comments/conclusions/advice. The relevant books for lawyers that I had read (and my discussions with some other "expert" psychologists) had forewarned me of some of the ethical issues that might arise if lawyers try to influence unduly the content of my "expert" reports (and any of my subsequent testimony).

Because I sometimes have had to make my position on this clear to lawyers from the outset, I have not often been requested to "amend" my reports' comments/conclusions/advice. However, a few years ago, a possibly difficult situation arose. Usually when lawyers contact me, I inform them that typically I will send to them my written report three weeks after my having received from them all of the relevant information. I also inform them, if necessary, that I will only be willing to receive their requests for clarity after they have received my written version.

One Thursday, a defense lawyer initially telephoned me to ask whether I would be willing to produce a report in a case involving a video-recorded interview by a police officer with an alleged child sexual abuse victim (who had said that it began with cuddles and progressed from there). Given that what he told me about the case suggested that my expertise was relevant, I agreed to produce a report. However, when I began to tell him about my three-week rule, he became agitated and told me that he needed a report before the criminal court case began on the following Monday. I said to him that this would not be possible. Nevertheless, he (skilfully) persuaded me to view the interview (and read the relevant case papers) on that Saturday. He said that he could quickly yet securely get these from his office in the north of England to my home on the south coast of England by asking his friend (who owned a small aeroplane) to fly him to my nearest airfield and from there take a taxi to my home. I told him that when he arrived mid-morning at my home, he would then need to go away for six hours to enable me not only to draft my report in my head but also to produce a finalized, written version. To this he rather reluctantly agreed.

Given that he was acting for the defense, I suspected that he would be seeking somehow to discredit the child's evidence/the skill of the interviewer. However, upon reading the relevant case papers and viewing/hearing the recorded interview several times, I came to the "expert" opinion that the interviewing was very much in line with the government's guidance and that the incriminating information provided by the child within the interview did seem to come from the child's memory. Thus, I expected myself to be placed in an ethical dilemma upon the defense lawyer's return. Therefore, in my mind I tried to prepare myself for what he would say upon reading my written report that I handed to him and that he immediately read. In preparing myself I sought to remember the ethical guidelines produced by relevant organizations of which I was a member (e.g., the British Psychological Society). However, I was not sufficiently prepared for what this defense lawyer said to me about this interview (which was the prosecution's main evidence). He said to me that earlier that week he had come to the same conclusion as I and that he now hoped that his client (i.e., the accused) would decide on Monday to take this lawyer's advice (which hitherto he had refused to do) and plead guilty!

One moral of this case study is that those behavioral and brain scientists who are willing to act as expert witnesses should not only prepare themselves to deal with ethical issues presented to them; they should also be aware of their own expectations and biases.

65 The Fragility of Truth in Expert Testimony

Phoebe C. Ellsworth

My research on topics related to psychology and law was inspired in large part by the belief that if judges are going to base their decision on assumptions about human behavior, as they often do, they should be informed by the best existing evidence from psychological science. Thus when I was first asked to be an expert witness, I regarded it as a great opportunity to provide the court with the best available empirical evidence on the question to be decided.

This turned out to be a lot more difficult than I had imagined. When sworn in, a witness is typically asked to swear to tell "the truth, the whole truth, and nothing but the truth." The whole truth is almost always out of the question. If I could provide a written summary reviewing the research on a fairly narrow question, such as how a prior identification of a person from a series of mug shots influences a person's later choice when faced with a lineup, I might be able to provide a pretty complete answer, although it would be full of qualifications and admissions that I just don't know about the effects of numerous third variables that have not yet been studied. Nonetheless, I would be satisfied that I had written a valid scientific summary.

But an expert is rarely asked for a scientific summary. Testimony takes the form of answers to questions. Any good attorney knows how to control what the expert says by asking questions that are likely to produce answers that forward his argument and avoiding questions that are not. If a relevant question is not asked, a bit of the whole truth is lost. There is not much the expert can do about this. The attorney will say, "Just answer the question, yes or no," and often the judge will join forces with the attorney and admonish the expert not to go beyond the question asked. If the expert insists on trying to add information that wasn't asked for, her career as an expert is likely to come to an abrupt end.

There are numerous systemic factors like this that prevent the expert from telling what she sees as the whole truth. But there are also subtler psychological influences that can undermine the truthfulness of what she actually does manage to say. Trials are adversarial, and even if she is

The Fragility of Truth

not paid, the expert almost inevitably comes to identify with the party that called her, to want to be a valuable member of the team. She gets her information about the case only from the lawyers on one side, she only gets to know the people on one side, and she comes to identify with them. And often the reason that she agreed to work with them is that she cares about the issue and believes that they are right. She knows that they are hoping that she will do well for them, and she doesn't want to let them down. But scientific accuracy is not the team's goal; winning is the goal.

If it is a good, egalitarian attorney-expert relationship, the expert will have collaborated in the preparation of the direct examination. She will have explained what she can say and told the attorney about areas where she does not believe that the scientific evidence supports the argument he wants to make. Usually the direct examination goes pretty smoothly. Sometimes, however, they attorney will not hold up his side of the agreement and will ask a question to which the expert cannot honestly give the answer the attorney is looking for, the answer that will help the side to win.

"Isn't it true, Dr. Ellsworth, that if there are three or more women on a jury, the jury will be more likely to acquit?" If asked that question outside the courtroom, it would be easy to say, "No, I have no reason to believe that's true." In court, however, this feels like a strike or a fumble, so the expert might say, "Yes, that sounds plausible," or "I don't know," or even simply "Yes."

Cross-examination is designed to undermine both the argument and the expert, so the pressure to give a confident, winning answer is even greater. The attorney may trot out a series of obscure references – "Are you familiar with the work of Osric? Voltemand? Marcellus?" After saying "no" three times and feeling like an ignoramus, when asked about the work of Yorick, whose name rings a faint bell, the relieved witness may respond with more confidence than is warranted. I have done this, and have said things that I wasn't really sure of because I didn't want to look or feel like a failure.

Could I be trapped into making the same mistake again? Probably. You really don't want to feel that it was your lame testimony that lost the case. What would I do differently? One move I've tried is to say, "No, I haven't heard of Osric's work. Can you tell me more about it?" The lawyer will probably say something like, "Osric found the opposite of what you're claiming." And then, of course, I ask follow-up questions about the research methods of the study, just as I would if I were talking to a colleague. It is my scientist's way of talking, and I'm comfortable. To the judge and the jury, my questions look like reasonable scientist questions. Typically they are *not* questions that the attorney is able or willing to answer, and he will generally drop that line of questioning, not only about Osric but about Voltemand and Marcellus as well.

The trick is to remember how you would talk in your familiar scientist role and not in your unfamiliar legal team member role, how you would talk in a context where you were being evaluated by other scientists – your orals, for example. In the courtroom it is easy to lose sight of the scientist role because there are no other scientists around to evaluate the integrity of the testimony. Everyone in the room is judging the witness's performance as a team member. Is she helping or hurting our side? None of your colleagues will ever know that you exaggerated or bent the truth. If the most distinguished psychologists in your field were in the courtroom, your testimony would probably be different. In one case in which I was involved long ago, we worried that the expert on the other side might say things that he would be embarrassed to say in front of respected experts in the field, so I invited Merrill Carlsmith and Lee Ross to visit the courtroom that day, making sure to introduce them to the witness before he testified. His testimony was considerably more cautious than we had been led to expect.

Nowadays, things have improved a little. The expert is now often asked for a written report, which allows her to provide an accurate summary of the scientific evidence on the question. This may not help her much when she in on the stand, as the attorneys still control the questioning and can pick and choose the parts of her report that they want to emphasize and the parts they want to exclude.

In sum, the expert is almost always forced into a situation of conflicting roles. The testimony that her side sees as a home run is usually not the testimony that her colleagues would admire. Like the participants in the Asch experiment, the witness usually gives the scientifically accurate answers, but sometimes succumbs to the temptation to compromise.

66 A Surprising Request from a Grant Monitor

Robert J. Sternberg

The business of obtaining, keeping, and renewing research grants is a stressful yet ongoing one. Given the times, those like me who have run large labs often have found themselves seeking multiple grants at once in order to make sure that they would not suddenly lose funding. At the peak of my research career (yes, that peak has passed!), I had a number of research grants, and felt on top of the world – until I received a strange request.

One day, I received a telephone call from the financial monitor on the largest of my grants, which was, to be precise, a subcontract from another university. I had had some contact with the individual before, but not a lot, since most of the financial reporting was through the primary recipient of the contract. When I had had contact with the monitors in DC, it usually was with the scientific monitor, not the financial one. The call started off pleasantly enough, but then turned to the real topic of the conversation. The monitor asked me whether I would be willing to help do the data analysis on his dissertation. I had not even realized that he did not yet have a doctorate. There was no discussion of exchange of money. The analysis was being requested as a personal, or perhaps professional, favor. He made the request, and there I was, sitting at my phone, sweating bricks. I told him that I certainly would think about it seriously and get back to him.

I thought seriously about his request. For one thing, I liked him well enough, and my personality is such that I try to help people out. I was quantitatively oriented and had no doubt that I, or at least one of the members of my research group, supervised by me, could do the data analysis. So my natural inclination was to help when asked. Moreover, he was the monitor for the finances of the grant, and of course I was more intent on pleasing him than I would have been of any random person who happened to call and ask for help with data analysis. Furthermore, the dissertation sounded reasonably interesting, so at least I would have some motivation to find out the results and perhaps learn something from them. So I had at least three reasons for wanting to help.

At the same time, the request struck me as odd. For one thing, I was not sure that the request was legitimate. Why wasn't he doing his own data analysis? I, as an advisor, would never allow a student to have someone else do data analysis for him or her. And I did not think that university policy would permit it either, whether the assistance was acknowledged or not (and I was pretty sure in this case it would fall into the "not" category). Moreover, the request had an unpleasant "tit for tat" character to it, somewhat as though it would be a situation in which I was helping him in exchange for him, as financial monitor, helping me. I was not sure I needed his help, but who could say for sure? Furthermore, I was concerned that my helping him perform the data analysis – which probably meant just doing the data analysis – would ethically compromise not only him but also me.

I called the principal investigator on the entire contract to ask his opinion of the situation. He told me that he thought the request was inappropriate and that I should not accede to it. But I was concerned that not helping the monitor might result in his somehow retaliating and making things unpleasant for me – and my lab. We were, at that point, dependent on the particular subcontract for a large number of jobs, and I was concerned that somehow the money would disappear.

I decided to call the scientific grant monitor, in confidence, to ask his advice. He worked for the contracting agency, and I thought he would be in the best position to tell me whether the request was legitimate or not. I thought I had a reasonably good relationship with him and would be able to trust him to keep my query in confidence and to give me good advice.

So I made the call, but things did not turn out as I had expected they would. He told me, first, that I should not help with the data analysis, but second, that he could not keep our phone call confidential – that he had a positive obligation to report my phone call and the request from the financial monitor that provoked my phone call to his supervisors. So, to my surprise, an event that I had hoped could be kept quiet did not stay quiet at all.

The upshot was that I did not help with the data analysis and did not hear again from the financial monitor about the request to help with his data analysis. I assume to this day that he was reprimanded, although I don't know that for sure. I did run into him again on one or two occasions, but neither of us mentioned the event; we both acted as though it never had happened, although our relations obviously were strained.

In retrospect, I believe the grant monitor's request was unethical, first because the monitor should have done his own data analysis, and second because he put me in a situation that would have created a conflict of interest. I believe that I probably did the right thing in not doing the data analysis and in reporting the contact, although I did not expect the scientific monitor to report my phone call. The incident made me particularly sensitive to the issue of creating conflicts of interest among those with whom I work. I never would want to put myself in the situation in which the financial monitor put me. And, to my knowledge, I never have put anyone in such a position.

67 Whoever Pays the Piper Calls the Tune: A Case of Documenting Funding Sources

Howard Tennen

The integrity of science requires transparency and freedom from interests other than the search for knowledge. A challenge to insulating the scientific enterprise from external influence is that the source of research funding tends to correlate with a study's findings and authors' interpretations of their findings. This "funding effect" (Krimsky, 2013) has been well documented, with a focus on the influence of for-profit/industry funders. Yet in psychological science, a comparable threat comes from ideological interests. This case study captures the potential influence of the ideological funding effect on psychological science.

The Ethical Challenge

Several years ago I was asked to serve as action editor for a journal article that reviewed and integrated several research areas. The authors' interpretation of the evidence and its implications for public policy was quite controversial. The reviewers took issue with several of the authors' interpretations, and in a revision, the authors addressed my concerns and those of the reviewers. Because there were several relatively minor issues that needed to be addressed in a second revision, I accepted the manuscript for publication pending these minor changes, but with the caveat that after accepting the final version, I would invite a commentary.

The final version of the manuscript was satisfactory. However, I noticed that this version included an acknowledgment of a source of funding that had not appeared in previous versions. As is common with research funding sources, the foundation's name revealed nothing about its mission. I discovered that this funding source was highly controversial, and that the controversy had been described in articles in the lay and scientific literatures. In response to my inquiry regarding the post-acceptance inclusion of the source of funding, the authors explained that it was an oversight. I myself had on at least one occasion forgot to list a source of funding, and

I had no reason to challenge the authors' explanation. Yet I believed that this situation presented an ethical challenge, not because the mission of the funding source concerned me personally and professionally (it did), but because the manuscript had been reviewed and accepted for publication without knowledge of this funding source. More to the point, the authors' controversial interpretations, and the public policy implications they put forward, were entirely consistent with the mission of the funding source, making plausible a "funding effect."

My Response to the Ethical Challenge

I took several steps before making my final decision to publish the manuscript. Before seeing the version of the manuscript that included the funding source, I had invited critical commentary to provide alternative interpretations and to challenge the policy implications offered by the authors of the target article. I discussed with the authors, and subsequently with the funding source (with the authors' consent), my concern that the authors' interpretation of the evidence cited in the manuscript was completely consistent with the mission of the funding source as it appeared on the funder's website. Not surprisingly, both the authors and the funder disagreed. I also corresponded with the editors of leading psychology journals, with several individuals in leadership roles in the journal that would publish the manuscript, and with close colleagues. Each of these individuals agreed with the plan to publish the manuscript with commentary, and two of them suggested several potential commentators. I consulted an ethicist and the journal's publisher, both of whom agreed with the plan. Finally, I consulted APA's *Ethical Principles of Psychologists*, which require that authors and editors avoid having scientific work used for other than scientific purposes. Having examined the funder's website, I was aware that the funder took the common step of citing articles based on research it had supported. I viewed such citations as implying an endorsement of the funder by the journal, and I insisted (knowing that my insistence might carry no weight) that the funder not cite this article on its website or in any of the funder's correspondence. The head of the funding source agreed in writing. Although publishing the article with commentary, receiving support for this decision from colleagues and experts I respected, and taking steps to adhere to APA's ethical principles allowed me to feel as though I was doing something, I was unable to address my key concern, which was that documentation of funding was not made available during the editorial review process.

Recent Literature as a Guide

During the years that have passed since this ethical encounter, a literature has emerged that offers evidence, opinion, and guidance related to the documentation of research funding. Evidence pointing to a funding effect has been documented in several areas of biomedical research. In response to this evidence, recommendations have been published by the International Committee of Medical Journal Editors, the Committee on Publication Ethics, the International Society of Addiction Journal Editors, the Center for Science in the Public Interest, and the Research Information Network (UK).

Babor and McGovern (2008) have suggested that in considering whether there is a conflict of interest, we should ask whether a reasonable person would feel misled if the situation or funding relationship were revealed only after the manuscript was published. From an action editor and reviewer's perspective, I would add that the situation should be revealed at the time of manuscript submission. Babor and McGovern go on to organize publishing misconduct in terms of Dante's seven circles of hell, with the first circle being carelessness, the sixth circle being plagiarism, and the seventh being data fabrication or falsification. They place failure to cite a funding source in the fourth circle, closer to plagiarism than to carelessness, and they conclude that all sources of funding need to be declared in a way that allows an average reader to recognize that there may be a potential conflict of interest. The Research Information Network recommends that in addition to acknowledging sources of funding, when research has not received external funding – a common occurrence in several areas of psychological science – a formal statement to this effect should in included in the publication.

Funding Documentation in Psychology

In search of models for reporting sources of funding, I informally reviewed the reporting practices of four psychology journals in the areas closest to my research interests, and two flagship psychology journals that publish research of general interest. In most of these journals, many of the articles published in the 2013 issues I selected did not include an acknowledgment of funding, and in none did the authors note that the research was not externally funded. From among these six journals, one now requires a formal declaration of conflicting interests and funding as part of the published article, and for unfunded research, a formal statement to that effect. In the two flagship journals, although all authors stated in a required declaration of conflicting interests that they had no conflicts of interest, none of the articles that did not list funding included a statement that the research was unfunded. This inconsistent pattern of reporting funding sources or their absence is in clear contrast to the reporting patterns I observed in the five leading journals in psychiatry, medicine, and neuroscience, where all fifteen journals require a statement of funding in the published article, and where every published article in the 2013 issues I examined listed funding sources or a statement that the research was not externally funded.

What I Would Do Differently with Hindsight

As psychological science moves increasingly into areas such as environmental issues, health, behavior genetics, and the causes of poverty and crime, there is an increased risk of ideological conflicts of interest. And as federal funding budgets dwindle and pay lines decline, investigators are turning to alternative funding sources or are conducting research without external funding. This situation underscores the importance and urgency of careful documentation of funding sources, or their absence.

Currently, when I serve as a manuscript's action editor, one of the first things I do is check the manuscript for documentation of funding sources, and request such documentation if it does not appear in the initial submission. I also research funding sources unfamiliar to me. However, I view this inefficient paper-by-paper approach as transitional. I am now working with a psychology journal publisher to follow the lead of journals in psychiatry, medicine, and neuroscience (and some journals in psychology) in requiring a statement regarding funding under a distinct heading in the manuscript. A side benefit of such a statement is that funding sources can be confident that their support will be documented in the scientific literature. Hopefully, all psychology journals will soon follow this approach. In the interim, I urge authors, editors, and publishers to remain vigilant regarding potential ideological conflicts of interest, and mindful of authors' ethical obligation to document all sources of funding when submitting their work for peer review.

REFERENCES

Babor, T. F., & McGovern, T. (2008). Dante's inferno: Seven deadly sins in scientific publishing and how to avoid them. In T. F. Babor, K. Stenius, S. Savva, & J. O'Reilly (Eds.), *Publishing addiction science: A guide for the perplexed* (2nd ed., pp. 153–171). Essex: Multi-Science Publishing Company.

Krimsky, S. (2013). Do financial conflicts of interest bias research? An inquiry into the "funding effect" hypothesis. *Science Technology Human Values*, 38, 566–587.

68 How to Protect Scientific Integrity under Social and Political Pressure: Applied Day-Care Research between Science and Policy

Marinus H. van Ijzendoorn and Harriet Vermeer

Introduction: The Thesis

Policy-relevant research should meet strict scientific criteria, especially when explosive political or social topics are involved. It is therefore desirable that decisions about funding of research proposals as well as supervision of testing the research hypotheses and the implementation of the research design always are the responsibility of an independent scientific forum. Obviously, policy makers and practitioners should be allowed to leave their mark on the formulation of the broad research question because they are the major participants of the discussion about the implications and applications of the research results. But applied scientific research is first and foremost aimed at the growth of knowledge and the search for truth, and should be executed independently of the interests of stakeholders, policy makers, and politicians; otherwise research and researchers risk the chance of being corrupted.

This thesis is illustrated by means of a concrete case study related to our research on quality of day care in the Netherlands. The House of Representatives initiated this study, which was executed by three research groups united in the Dutch Consortium for Research into Child Care (NCKO) (Nederlands Consortium Kinderopvang Onderzoek, 2005).[1] Ethically responsible participation in the planned follow-up study of the NCKO became impossible because of a boycott by one of the most powerful stakeholders in the domain of day care, and by the ambiguous role of the ministerial authority that commissioned the research. The Leiden research team refused to bow to the social and political pressures and left the consortium, leaving behind several million dollars of grant money.

Scientifically valuable applied research should be protected against interference by interest groups through independent funding and careful separation of responsibilities between the subsidizing agency, the research team, an "advisory board," and the scientific forum.

The Case Study: Day-Care Quality

One of the major challenges of the Dutch Childcare Act, which came into force in 2005, is to enable parents to combine work and family. The legislature explicitly values hygiene and physical safety in day care, but also adequate stimulation of children's development and the promotion of their social-emotional security. According to the law, day care should provide more than mere custodial care.

Following this description of basic goals and methods of day care, the NCKO conducted a nationwide quality assessment of day-care centers in 2005. This assessment was the third national assessment of quality of day care for 0-to-4-year-olds since 1995, and was initiated and subsidized by the Dutch Ministry of Social Affairs and Employment (SZW). In all three assessments, quality was examined in 40 to 50 different day-care centers, based on several hours of detailed, structured observation in each center. For this purpose we used internationally established and standardized measures that cover the descriptions and goals of day care, as enacted in the Dutch Childcare Act.

The assessment unfortunately showed a low level of quality of Dutch day-care centers. Average quality ratings – on a scale of 1 to 7 – were slightly higher than 3, distinguishing between "poor" (lower than 3) and "moderate" (from 3 to 5) to 'high' (5 or higher) quality. Childcare quality had significantly declined across the preceding decade. Whereas in 1995 no center had received the label "poor," 6% of the centers were of poor quality in 2001, but in the 2005 a formidable 36% of the centers received this poor grade (using the same criteria). Results from international studies using the same instruments showed that the quality of Dutch day-care lagged behind day-care quality in countries such as Germany, Canada, and the United States.

Public Panic and Political Pressure

The results of the NCKO study were extensively discussed in the national media and gave rise to much commotion in day-care practice. Also, stakeholders and politicians were concerned, and the minister of SZW had to justify the unsatisfactory quality of day care to the House of Representatives. The minister of SZW took the initiative to commission a follow-up study, a fourth and more extensive national quality assessment. The NCKO was commissioned to conduct this research, which would start in 2006. Meanwhile, however, one of the national interest groups in day care called for a boycott of the NCKO on its website.

This entrepreneurial stakeholder was dissatisfied with the earlier study outcomes and advised its affiliated day-care centers to close the gates to NCKO researchers. The effect of this boycott was immediately noticeable in refusals of day-care centers to participate in the follow-up study. The stakeholder received a ministerial reprimand for this boycott call but remained present in the scientific advisory board of the NCKO. The boycott of the NCKO was continued, despite the insistence of the minister of SZW to stop it. In the meantime, the same ministry put more pressure on the NCKO scientists to interrupt the ongoing study, to submit a new research proposal accommodating some of the wishes of the stakeholder, and to resume the study only after a year. After some heated debates with the authorities that commissioned the study and after extensive discussions with the other research teams, the Leiden researchers decided to leave the consortium, which had just received several million dollars of funding.

What to Do?

The problem is obvious. Public officials, most often not scientifically trained, formulate research assignments to which commercial and university research groups can subscribe. This is an outstanding example of political-administrative influence as a "fourth power" on the direction of scientific research. The selection of the winning research proposal ultimately lies in the hands of the ministry, not the scientific forum, and fierce competition leads to "murderous contracts," not only in financial terms but also in detailed prescriptions of methods and outcomes of a study by stakeholders. Conditional rights of publication of the research results may even be enforced, although no danger for national security is at stake. Interest groups intrude on the supervision of the research process and play a leading role without any scientific expertise, not only at the stage of proposing research questions but also in the subsequent stages of the research process. Disappointing or unexpected results may trigger unwanted media attention and cause panic among policy makers and stakeholders. The researchers are blamed for the message.

Some measures may prevent such problems, or at least reduce them. First, policy-relevant research should be scientifically grounded, and only after the research design has been approved by the scientific forum should the question of relevance become an issue. Validity and reliability of the research design and methods are necessary conditions for social or political relevance. The scientific forum should therefore get a central role in applied as well as in basic research.

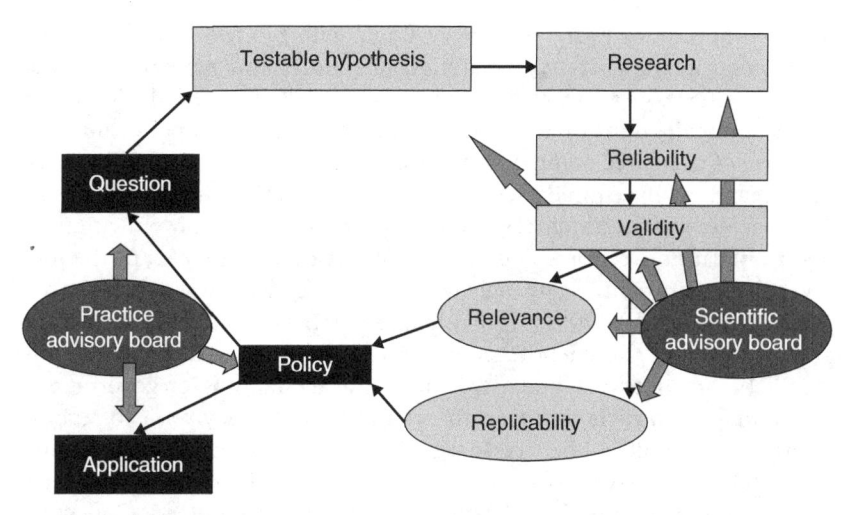

Figure 68.1 Research cycle from practice to theory and back again

Second, ministries and stakeholders may outline the broad research question, paving the way for *problem-driven* or *concern-driven* research, driven by interests other than just scientific goals. But scientists should determine the next crucial steps toward theory, hypotheses, and methods. The implication of this principle is that there should be a distinction between a "practice advisory board" and a "scientific advisory board," in which the practice advisory board has input in the discussion about research questions and practical implications, whereas the scientific advisory board provides feedback on the scientific stages of the study (see Figure 68.1). The distinction in expertise is clear; what should be guaranteed is an unambiguous distinction and separation of power and corresponding responsibilities.

A clear demarcation of responsibilities, rights, and duties of all parties is needed in the various stages of research. The research cycle begins and ends with a discussion between researchers and the practice advisory board in which both the broad research questions and the practical implications of the results – as suggested by policy makers – are subject of consensus. Between head and tail of the research cycle researchers are only accountable to the forum of scientists with ample expertise in the relevant area of research.

Conclusion

The case of the day-care quality study is obviously an extreme case but not a unique one (Köbben & Tromp, 2003). The case is at the far end of a continuum, or rather a slippery slope, with too few built-in brakes. The proposal for a stricter separation of powers and responsibilities is usually unnecessary but at the same time is a badly needed protection against ethically unwanted influences of stakeholders and policy makers. Because of the fierce competition for scarce scientific resources, individual researchers have hired themselves out as "scientific mercenaries" and have agreed to sign unfair research contracts, sometimes even with limited publication freedom. Their scientific integrity is at stake, as the formal agreements contrast strongly with universally accepted guidelines for researchers (KNAW, 2005). Individual researchers should feel supported and protected by the scientific community. A first step to collective action would be an inventory and (legal and ethical) analysis of governmental standard contracts that researchers must sign if they want to receive grants to conduct applied research. Professional scientific associations (e.g., APA, APS, SRCD) should design standard agreements in which separation of power and publication freedom are guaranteed. Collective pursuit of such guarantees for academic freedom of applied researchers is urgently needed. Without the support of the scientific community, the price to pay for scientific integrity may be too high for the individual researcher.

NOTE

1 The Dutch Consortium for Research into Child Care (NCKO) was, until April 2006, a collaboration between educationalists and developmental psychologists of the University of Amsterdam, Leiden University, and the Radboud University Nijmegen. After that the Leiden research group decided to leave the consortium. The described research was funded by the Ministry of Social Affairs and Employment (SZW).

REFERENCES

KNAW-Werkgroep Opdrachtonderzoek. (2005). *Wetenschap op bestelling. Over de omgang tussen wetenschappelijke onderzoekers en hun opdrachtgevers* [*Science on command: About the contact between researchers and their clients*]. Amsterdam: KNAW.

Köbben, A. J. F., & Tromp, H. (2003). *De onwelkome boodschap, of hoe de vrijheid van wetenschap bedreigd wordt.* [*The unwelcome message or how the freedom of science is threatened*]. Amsterdam: Mets & Schilt Uitgevers.

Nederlands Consortium Kinderopvang Onderzoek. (2005). *Kwaliteit van Nederlandse kinderdagverblijven: Trends in kwaliteit in de jaren 1995–2005* [*Quality of Dutch daycare centers: Trends in quality over the years 1995–2005*]. Amsterdam/Leiden/Nijmegen: NCKO.

69 Commentary to Part XI

Susan T. Fiske

The usual prescription for conflict of interest, including financial gain, is disclosure, and certainly transparency is a necessary foundation for ethical processes. But psychology suggests disclosure will not suffice. Actors who have a personal conflict of interest, being human, will discount the extent to which they are influenced by personal gain. Observers, even knowing about the conflict of interest, will not necessarily be impartial judges of the effects on the actors. The best goal is no conflict of interest, but we do not live in a perfect world. So reflection and discussion will have to serve.

Epilogue: Why Is Ethical Behavior Challenging? A Model of Ethical Reasoning

Robert J. Sternberg

A question one might ask upon reading the essays in this book is why, at least for some people, ethical behavior is so challenging. Drawing in part on Latané-Darley's (1970) model of bystander intervention, I have constructed a model of ethical behavior that would seem to apply to a variety of ethical problems. The model specifies the specific skills students need to reason and then behave ethically.

The basic premise of the model is that ethical behavior is far harder to display than one would expect simply on the basis of what we learn from our parents, from school, and from our religious training (Sternberg, 2009a, 2009b, 2009c). To intervene in an ethically challenging situation, individuals must go through a series of steps, and unless all of the steps are completed, the individuals are not likely to behave in an ethical way, regardless of the amount of training they have received in ethics, and regardless of their levels of other types of skills. The example I draw on throughout this brief essay is one of questionably ethical behavior by a professor who is the director of a federally funded laboratory operating in the context of a university.

According to the proposed model, enacting ethical behavior is much harder than it would appear to be because it involves multiple, largely sequential steps. To behave ethically, the individual has to

1. recognize that there is an event to which to react;
2. define the event as having an ethical dimension;
3. decide that the ethical dimension is of sufficient significance to merit an ethics-guided response;
4. take responsibility for generating an ethical solution to the problem;
5. figure out what abstract ethical rule(s) might apply to the problem;
6. decide how these abstract ethical rules actually apply to the problem so as to suggest a concrete solution;
7. prepare for possible repercussions of having acted in what one considers an ethical manner; and
8. act.

Consider each step in turn, first in response to a professor who requires that each paper coming out of his federally funded university lab has his name on it as a co-author, and second in response to ethical situations described previously in this book.

Recognize That There Is an Event to Which to React

In cases where there has been an ethical transgression, the transgressors often go out of their way to hide that there is even an event to which to react.

Sometimes, professors who commit ethical transgressions do not even recognize their own breaches. For example, they may put their name as a coauthor on every paper that comes out of their lab, even if they had absolutely nothing substantive to do with the work (i.e., no role in funding, conceptualizing, executing, or writing up the work). Their sole role is providing the lab facilities and lab environment. They may feel justified because the work was done in their lab. Moreover, it may be that few people or no one in the lab questions the practice because each member of the lab assumes that the professor's name on every paper is standard practice.

Suppose, in this instance, Professor Smith puts his name on every paper coming out of his lab, regardless of his contribution. But perhaps someone, Dr. Jones, a postdoc, believes that something is wrong. He recognizes a situation that just does not seem right to him. Then what?

Consider further the case of the "serial collaborator" (Geary, Chapter 37). You discover that a collaborator who offers his services to one team after another has, in all likelihood, faked data. You decide not to use any data he collected. Case closed. Or is it? Is that the end of the situation, or, as Geary points out, is there more to it? Do you have to identify exactly to whom he reports, and to what authority one should report his misdeeds? What reach does one have in another country on another continent? The Geary case points out that sometimes it is difficult not only to recognize that there is an event to which to react but also to figure out just what the full scope of the event is.

Define the Event as Having an Ethical Dimension

Given that one acknowledges that there is an event to which to pay attention, one still needs to define it as having an ethical dimension. Given that perpetrators will go out of their way to define the situation otherwise – as a nonevent, an internal conflict that is no one else's business, and so on – one must actually redefine the situation to realize that an ethical component is involved.

In the case of Professor Smith, if Dr. Jones (or someone else) has doubts, he must ask whether indeed this is a question of ethics or of something else. For example, maybe the professor is acting ethically but is overestimating his contribution to the work – for example, he may feel that the use of equipment he has bought from a grant is sufficient for his name to be on the paper, or that the unique configuration of his lab offers benefits that make his contribution important. Dr. Jones may decide that Professor Smith has used less than excellent judgment but is not, in fact, unethical. Or he may decide that Professor Smith is something of a megalomaniac, but that the issue is one of personality rather than of ethics.

Peter Lovibond, in Chapter 18, discusses the situation in which a student works on a project – perhaps a dissertation – and then disappears without publishing it. A similar point is made in Chapter 20 by Roberts, Beals-Erickson, Evans, Odar, and Canter with regard to dissertations in particular. From the student's point of view, there may be no ethical issue at all. After all, she finished the project or dissertation and moved on. To the extent that her research was funded, she may see her ethical obligation as discharged. But as Lovibond and Roberts and colleagues point out, the advisor and others may see the ethical obligation as over only when the data are published, because dissertations are rarely read by individuals outside the dissertation committee, and hence the science has not truly become public. Lovibond's and Roberts and colleagues' essays show the extent to which people may differ in what they view as a relevant ethical dimension. A good solution, as suggested, is to have students sign some kind of agreement regarding publication before they even undertake the dissertation research.

Decide That the Ethical Dimension Is Significant

If one observes a driver going one mile per hour over the speed limit on a highway, one is unlikely to become perturbed about the unethical behavior of the driver, especially if the driver is oneself. A genocide is a far cry from driving one mile per hour over the speed limit. And yet, if one is being told by cynical, dishonest leaders that the events that are transpiring are the unfortunate kinds of events that happen in all countries, then it may not occur to people that the event is much more serious than its perpetrators are alleging it to be.

In the case of Professor Smith, perhaps Dr. Jones comes to the conclusion that the behavior of the professor is unethical. But he may decide that it is not unethical at a level that requires any further intervention. For example, Dr. Jones may decide that although Professor Smith's name should not be on some of the papers, at least all of those who have made significant contributions do have their names on the relevant papers. As a result, Dr. Jones may decide that the matter is just not worth pursuing. Or he may suspect that Professor Smith is not the most ethical professor in the world, but who is he, Jones, to judge?

People often differ in what they see as significant when it comes to ethical reasoning and decision making. William Buskist discusses in Chapter 5 the issue of extra credit for students who want to improve their grade. He suggests that offering some students an extra-credit option without offering the same option to other students is unfair. I agree. But what for him (and others) is a matter of principle may not be quite the same for others. Other professors may feel that one must decide cases individually, offering extra credit to individual students in cases that are unusual (such as a student who has experienced a personal tragedy and is trying to avoid, as a consequence, failing or otherwise doing poorly in a course.) We all need to pay serious attention to the issue of whether something we see as not all that significant may in fact be significant, or vice versa.

Take Personal Responsibility for Generating an Ethical Solution to the Problem

People may allow leaders to commit wretched acts, including genocide, because they figure it is the leaders' responsibility to determine the ethical dimensions of their actions. Isn't that why they are leaders in the first place? Or people may assume that the leaders, especially if they are religious leaders, are in a uniquely good position to determine what is ethical.

Suppose, in the case of Professor Smith, that Dr. Jones decides that the behavior of Professor Smith is unethical. The issue now confronting him is what to do about it. He may decide that someone should do something about it, but not him. This might be the case especially if, on Dr. Jones's particular papers, Professor Smith always has made at least some substantive contribution.

The issue of personal responsibility figures prominently in the discussion of Kathy Hirsh-Pasek and Marsha Weinraub (Chapter 47) of when investigators have a responsibility to tell parents that their child may have a developmental disability or other problem. If children are research subjects, typically our responsibility as investigators is to debrief parents regarding the general results of the research we have done, not to debrief them regarding their own children's performance. In some cases, it clearly would be unethical to debrief parents regarding their children's performance, as when, say, adolescents are guaranteed confidentiality in interviews. Hirsh-Pasek and Weinraub dealt with this problem by setting essentially statistical criteria for debriefing parents and then took great care in the nature of their debriefing about the individual child. We all, as investigators, need to address our personal responsibility as the result of our conduct of research.

Figure Out What Abstract Ethical Rule(s) Might Apply to the Problem

Most of us have learned, in one way or another, ethical rules we are supposed to apply to our lives. For example, we are supposed to be honest. But who among us can say he or she has not lied at some time, perhaps with the excuse that we were protecting someone else's feelings? By doing so, we insulate ourselves from the effects of our behavior.

Dr. Jones may decide that Professor Smith's behavior is unethical, is significant, and that he, Dr. Jones, has a responsibility to do something. Now he must ascertain exactly what rule is being broken, if any. Dr. Jones probably cannot go to the bible, or any standard book of ethics, to figure out the relevant rule. The problem is too specific to professional disciplines. Dr. Jones probably will need to go to a source within his own profession, such as the American Psychological Association's *Code of Ethics* or the *APA Publication Manual*. Dr. Jones then would have to find out what exactly he should do in this particular instance.

The problem of deciding what abstract ethical rule applies in psychological research comes out clearly in Diane Halpern's essay (Chapter 14). When you use someone else's scale in research in order to give the research greater credibility, what are your responsibilities to the creator(s) of the scale? Do you have to inform them of your use of the scale? Do you have to use the scale in the exact form the creator(s) used it? Is it unethical to reprint the scale so that others can determine its content validity (especially if it previously has been published)? The problem is that we cannot possibly learn in graduate school or at any other single time period all of the ethical rules that apply in research, so we need constantly to be learning and also using our common sense to ensure we act in ways that are ethical and also help advance our science.

Decide How These Abstract Ethical Rules Actually Apply to the Problem so as to Suggest a Concrete Solution

This kind of translation is, I believe, nontrivial. It is often hard to figure out exactly how to apply a rule to a particular situation. For example, one may detect a genocide halfway around the world, know one should act, but have no clue as to just what one should do.

In the case of Dr. Jones, Jones may find in consulting sources that professors should put their names on papers only if they have made a substantive contribution to the work in the paper – such as contributing to the design, execution, or writing up of the paper. Dr. Jones might find that simply being the principal investigator in a lab is not sufficient. But applying the rule may not be entirely straightforward, because it is likely that Professor Smith's interpretation of his contribution to a given paper may be different from Dr. Jones's interpretation. In this case, Dr. Jones may find himself in a serious conflict with Professor Smith, leading to the next issue. How does Dr. Jones know that his interpretation of the situation is correct or, even if it is, that it will prevail in a hearing of some kind?

In Chapter 45, Haslam, Reicher, and McDermott show just how far one can go in ensuring that one has not only an ethical basis for one's research but also a way of implementing it. They point to the Zimbardo Stanford Prison Experiment as one of the great empirical demonstrations of all times, but also as one, like the Milgram experiments on obedience, that never would have passed muster with modern institutional review boards. Many people would have concluded that there just would be no way to do something like the Stanford Prison Experiment in an ethical way according to contemporary standards, but the authors show how even such a fraught demonstration can be done in an ethical way if one devises ingenious and comprehensive protections for the subjects involved in the study.

Prepare for Possible Repercussions of Having Acted in What One Considers an Ethical Manner

When Harry Markopolos (see Markopolos, 2011) pointed out to regulators that Bernard Madoff's investment returns had to be fraudulent, no one wanted to listen. It was Markopolos who was branded as a problem, not Madoff. In general, when people blow the whistle, they need to be prepared for their bona fides to be questioned, not necessarily those of the person on whom they blew the whistle.

In the case of Dr. Jones, if he does indeed act against Dr. Smith, he needs to be prepared for possible repercussions. For example, if Jones brings a complaint against Professor Smith and Smith wins with the dean or other arbiter, Dr. Jones may find himself out of the lab and likely out of a job. Or Dr. Jones may win, but then find that Professor Smith is only reprimanded but still is in control of his lab. In this instance,

Dr. Jones may have won the battle but lost the war – he is still out of a job. Or he may be allowed to stay in the lab but find that he will get absolutely no support from Professor Smith when it the time comes to find an assistant professorship. The potential costs of acting are great; the benefits – uncertain.

The issue of negative repercussions comes out particularly clearly in Chapter 21 by Weisstein. The author was a junior investigator, and senior investigators, prominent in their field and also journal editors, wanted to rerun her experiments in a supposedly "better" way. What is one to do in the face of what appears to be idea theft by investigators much more senior than oneself, who potentially have the power to make one's ascent in the field difficult, or even to thwart it altogether? Weisstein handled the situation by submitting quickly to a different journal, but in some cases, the senior individuals are one's supervisor or senior colleagues in one's own department and may later vote against one in a tenure or promotion decision (see also Chapter 12 by Beutler), or fail to support one for a job in the first place. In these cases, such situations are particularly vexing.

Act

In ethical reasoning as in creativity, there may be a large gap between thought and action. Both often involve defying the crowd, and hence even people who believe a certain course of action to be correct may not follow through on it.

In Latané and Darley's (1970) work, the more bystanders there were, the less likely an individual was to take action to intervene. Why? Because one figured that, if something was really wrong, then someone among all the others witnessing the event would have taken responsibility. You are better off having a breakdown on a somewhat lonely country road than on a busy highway, because a driver passing by on the country road may feel that he or she is your only hope.

In the case of Professor Smith and Dr. Jones, Jones may reach the point where he believes that he is willing to take the risks of reporting what he views as unethical behavior by Professor Smith. But when it comes to acting, he may find himself paralyzed or simply delaying doing anything. Over time, he simply may fail to act, even though he believes that he should act. In the end, he may do nothing.

One of the harder situations in which to act appropriately is that of one's being confronted with student plagiarism (see Chapter 3 by Plous). On the one hand, one probably knows that one should report it to a central authority, if only to prevent the student from repeating the act of plagiarism in other classes without the benefit of the instructor's finding out about the plagiarist's past history. On the other hand, few professors want to get involved in student judicial proceedings and even the possibility of a lawsuit by the student or his parents. In some situations, we know what we have to do but may stop short of action because the action may become so burdensome.

Conclusion

Deciding how to confront ethical challenges is one of the biggest challenges we will face in our lives (Rogerson, Gottlieb, Handelsman, Knapp, & Younggren, 2011; Sternberg, 2011a, 2011b). In the behavioral and brain sciences, as the essays in this book have shown, there are numerous kinds of challenges to acting ethically in challenging situations. Sometimes the ethical lapses are so egregious that everyone sees them, and action is easy. But at least as often, one risks confronting shades of gray. What does one do then? That is the question the essays in this book have addressed. Hopefully, they will have helped our readers decide how to handle the difficult ethical challenges they face in their own careers. The model described here can be applied to each of the ethical challenges reported in the chapters of this book. But it is a model of ethical reasoning, not of normative solutions. It remains for all of us to use our ethical reasoning to come up with fair, just, and reasoned solutions.

REFERENCES

Latané, B., & Darley, J. (1970). *The unresponsive bystander: Why doesn't he help?* Englewood Cliffs, NJ: Prentice-Hall.

Markopolos, H. (2011). *No one would listen: A true financial thriller.* New York: Wiley.

Rogerson, M. D., Gottlieb, M. C., Handelsman, M. M., Knapp, S., & Younggren, J. (2011). Nonrational processes in ethical decision making. *American Psychologist*, 66(7), 614–623.

Sternberg, R. J. (2009a). Ethics and giftedness. *High Ability Studies*, 20, 121–130.

(2009b). A new model for teaching ethical behavior. *Chronicle of Higher Education*, 55(33), B14–B15.

(2009c). Reflections on ethical leadership. In D. Ambrose & T. Cross (Eds.), *Morality, ethics, and gifted minds* (pp. 19–28). New York: Springer.

(2011a). Ethics: From thought to action. *Educational Leadership*, 68(6), 34–39.

(2011b). Slip-sliding away, down the ethical slope. *Chronicle of Higher Education*, 57(19), A23.

Index

academic advisors, 48–49
academic dishonesty, 4, 114–118
Amazon, 153
American Association for the Advancement of Science (AAAS), 45–46, 188
American Beverage Association (ABA), 198
American Journal of Psychology, 35
American Psychiatric Association (APA), 36
American Psychological Association (APA), 36, 42, 61, 186, 209
Association for Psychological Sciences (APS), 112
attribution, 8
authorship, credit, 35–37, 38–40, 50.
  See also ordering of authorship

BBC Prison Study
  ethical challenges, resolution, 135–138
  social identity model of tyranny and, 134–135
Belmont Report, 148, 157
BlackBerry Project, 76–78
Brilliant Brothers (Livio), 89
Byrd, James, 155

Cain, Daylian, 199
Canadian Psychological Association, 174
case studies, overview
  five key issues, xvii
  importance of, xvi
cheater-detection, 11
children
  behavior observation of, 140–142
  child care studies, 145–148
  eyewitness abilities of, 131–133
  temptation probe technique and, 140–143
Children's Hospital of Philadelphia, 198
Chronicle of Higher Education, xvi

claiming ownership, 66–67
Clark-Pine, Dora, xv
collaboration, 5–6, 111–112, 113, 128
Committee on Publication Ethics (COPE), Ethical Guidelines for Peer Reviewers, 46
competition, xvi
confidentiality limits, 76. *See also* digital communication, adolescents, confidentiality
courses of action, 174–176
Crabbe, John, 106–107
culture, 161–163, 177

data analysis, motivated reasoning, 87–88
data fabrication, 122–123, 124
data patterns, 124
data sharing policies, 92–93
data, theory matching, 85–86
digital communication, adolescents, confidentiality, 76
disgruntled colleagues, 20
dissertations, non-publication, 59–61
Dutch Childcare Act, 213–215
Dutch Consortium for Research into Child Care (NCKO), 212–214

East-West Center, 161–162
ego issues, 177
Ethical Guidelines for Peer Reviewers (Committee on Publication Ethics), 46
Ethical Guidelines for Reviewers (American Association for the Advancement of Science), 45–46
Ethical Principles of Psychologists and Code of Conduct (American Psychology Association), 61, 167, 209
ethical reasoning model, 219–220
  abstract ethical rules, concrete solutions, 223–224
  applicable abstract ethical rules and, 223

227

ethical reasoning model (*cont.*)
  ethical dimension significance, 221–222
  event ethical dimension, 220–221
  event to which to react, 220
  personal responsibility to generate ethical solution and, 222
  repercussions preparation, 224–225
  taking action and, 225
exam cheating, plagiarism *vs.*, 9
exam proctoring, 6–7
experimental design
  design matters, 101–102
  replacement, refinement, reduction and, 101–102
  results, consequences, 102–103
expert testimony
  ethical issues, personal expectations and bias, 200–201
  truth, scientific summary and, 202–204
extra credit
  emotional appeals for, 15
  fair, equal use of, 15–17

fair evaluation, deservedness issue, 28–29
fairness doctrine, 22, 31
Family Education Rights and Privacy Act (FERPA), 21
Federation of Associations in Behavioral and Brain Sciences, xvii
fidelity, responsibility in leadership, 167–170
file drawer problem, 60

grading policies
  fairness doctrine and, 22
  subjective element, 23–24

HARKing, 94–96

idea-poaching, 44–45
industry funding. *See* research funding
informed consent, 120
International Positive Psychology Association, 174
International Society for Prevention of Child Abuse and Neglect (ISPCAN), 149

Journal of Mathematical Psychology, 94

Livio, Mario, 89
loyalty, 177

matching data sets, 126
Mechanical Turk, 153
moral education, 55

National Institute of Child Health and Human Development (NICHD), 145
National Research Ethics Committee, 150
NCKO (Dutch Consortium for Research into Child Care), 212–214

Obama, Barack, 85
ordering of authorship, 50–51
owning errors, 89–90

peer-review, 44, 47
personal networking, 163
plagiarism
  defined, 10
  detection of, 10–11
  electronic communication, Internet and, xvi
  exam cheating *vs.*, 9
  from external sources, xvi
  prevalence of, xv
privileged documents, 44–45
professional responsibility, 7
Psychological Bulletin, 186, 187
published scales, 41–42

Quebec Society for Research in Psychology, 174

replaceability, 51–52
research funding
  blindspots, self-forgiveness and, 198
  critics of, 197
  disclosure of potential conflicts and, 199
  Dutch Childcare Act, 213–215
  funding documentation in psychology, 210–211
  funding effect, 208
  funding sources documentation, 208–209
  monitoring of, 197
  motives, 197
  recent literature as guide, 210
  research grants monitoring, 205–207
research grants monitoring, 205–207
Research Information Network, 210
research relationships, power dynamics, 55–56
reviewing, editing of manuscripts
  bias in review process, 183–184
  dissenting experts, published commentaries opportunities, 189
  hot button issues, 189
  qualified reviewers, 189
  readers' comments, ongoing forum for issues, 189–190

repeat reviewing, 181–182
third parties, conflicts of interest, 191–193
Rind et al. affair, 186–190

Science Magazine, 64
scientific data, withholding, 29–30
sexual violence victims, research ethics, 149–152
Slacker Stack, 61
Standard Ethical Principles of Psychologists and Code of Conduct (APA), 45
Stanford Prison Experiment (SPE), 134
Stormfront, 155–156
student contact, unpublished research dilemma, 53–54
student requests
  evidence, documentation for, 18–19
  future professions and, 25, 27
  incomplete grades and, 26

stalking, harassment and, 25
student's disengagement, irresponsibility and, 25, 27
system-level monitoring, 11

tenure, 165–166, 171–173
True Score ANCOVA, 83
truth telling, competing motives, 57–58
Turnitin.com, xvii, 10

unreliable data, 120

Wesleyan University Honor Code, 9–10
White Aryan Resistance, 155
withholding scientific data, 29–30
women, feminism, 65
World Church of the Creator, 155
World Health Organization (WHO), 149
writing assignments, guidelines, 9

Zimbardo, P., 134–135